Applications of Big Data Analytics

Mohammed M. Alani • Hissam Tawfik
Mohammed Saeed • Obinna Anya

Editors

Applications of Big Data Analytics

Trends, Issues, and Challenges

Springer

Editors
Mohammed M. Alani
Al Khawarizmi International College
Abu Dhabi, UAE

Hissam Tawfik
Leeds Beckett University
Leeds, UK

Mohammed Saeed
University of Modern Sciences
Dubai, UAE

Obinna Anya
IBM Research
San Jose, CA, USA

ISBN 978-3-030-09497-3   ISBN 978-3-319-76472-6   (eBook)
https://doi.org/10.1007/978-3-319-76472-6

This Springer imprint is published by the registered company Springer International Publishing AG part of Springer Nature.
The registered company address is: Gewerbestrasse 11, 6330 Cham, Switzerland

Preface

Big Data comes in high volume, velocity, and veracity, and from myriad sources, including log files, social media, apps, IoT, text, video, image, GPS, RFID, and smart cards. The process of storing and analyzing such data exceeds the capabilities of traditional database management systems and methods, and has given rise to a wide range of new technologies, platforms, and services—referred to as Big Data Analytics. Although the potential value of Big Data is enormous, the process and applications of Big Data Analytics have raised significant concerns and challenges across scientific, social science, and business communities.

This book presents the current progress on challenges related to applications of Big Data Analytics by focusing on practical issues and concerns, such as the practical applications of predictive and prescriptive analytics especially in the health and disaster management domains, system design, reliability, energy efficiency considerations, and data management and visualization. The book is the state-of-the-art reference discussing progress made and problems encountered in applications of Big Data Analytics, as well as prompting future directions on the theories, methods, standards, and strategies necessary to improve the process and practice of Big Data Analytics.

The book comprises 10 self-contained and refereed chapters written by leading international researchers. The chapters are research-informed and written in a way that highlights the practical experience of the contributors, while remaining accessible and understandable to various audiences. The chapters provide readers with detailed analysis of existing trends for storing and analyzing Big Data, as well as the technical, scientific, and organizational challenges inherent in current approaches and systems through demonstrating and discussing real-world examples across a wide range of application areas, including healthcare, education, and disaster management. In addition, the book discusses, typically from an application-oriented perspective, advances in data science, including techniques for Big Data collection, searching, analysis, and knowledge discovery.

The book is intended for researchers, academics, data scientists, and business professionals as a valuable resource and reference for the planning, designing, and implementation of Big Data Analytics projects.

Organization of the Book

The chapters of the book are ordered such that chapters focusing on the same or similar application domain or challenge appear consecutively. Each chapter examines a particular Big Data Analytics application focusing on the trends, issues, and relevant technical challenges.

Chapter 1 discusses how recent innovations in mobile technologies and advancements in network communication domain have resulted in the emergence of smart system applications, in support of the wide range and coverage provision, low costs, and high mobility. 5G mobile network standards represent a promising cellular technology to provision the future of smart systems data traffic. Over the last few years, smart devices, such as smartphones, smart machines, and intelligent vehicles communication, have seen exponential growth over mobile networks, which resulted in the need to increase the capacity due to generating higher data rates. These mobile networks are expected to face "Big Data" related challenges, such as explosion in data traffic, storage of big data, and the future of smart devices with various Quality of Service (QoS) requirements. The chapter includes a theoretical and conceptual background on the data traffic models over different mobile network generations and the overall implications of the data size on the network carrier.

Chapter 2 explores the challenges, opportunities, and methods, required to leverage the potentiality of employing Big Data into the assessing and predicting the risk of flooding. Among the various natural calamities, flood is considered one of the most frequently occurring and catastrophic natural hazards. During flooding, crisis response teams need to take relatively quick decisions based on huge amount of incomplete and, sometimes, inaccurate information mainly coming from three major sources: people, machines, and organizations. Big Data technologies can play a major role in monitoring and determining potential risk areas of flooding in real time. This could be achieved by analyzing and processing sensor data streams coming from various sources as well as data collected from other sources such as Twitter, Facebook, satellites, and also from disaster organizations of a country by using Big Data technologies.

Chapter 3 discusses artificial intelligence methods that have been successfully applied to monitor the safety of nuclear power plants (NPPs). One major safety issue of an NPP is the loss of a coolant accident (LOCA), which is caused by the occurrence of a large break in the inlet headers (IH) of a nuclear reactor. The chapter proposes a neural network (NN) design methodology in three stages to detect the break sizes of the IHs of an NPP. The results show that the proposed methodology outperformed the MLP of the previous work. Compared with exhaustive training of all two-hidden layer architectures, the speed of the proposed methodology is faster than that of exhaustive training. Additionally, the optimized two-hidden-layer MLP of the proposed methodology has a similar performance to exhausting training. In essence, this chapter is an example of an engineering application of predictive data analytics for which "well-tuned" neural networks are used as the primary tool.

Chapter 4 discusses a Big Data Analytics application for disaster management leveraging IoT and Big data. In this chapter, the authors propose the use of drones or Unmanned Aerial Vehicles (UAVs), in a disaster situation as access points to form an ad hoc mesh multi-UAV network that provides communication services to ground nodes. Since the UAVs are the first components to arrive at a given disaster site, finding the best positions of the UAVs is both important and non-trivial. The deployment of the UAV network and its adaption or fine-tuning to the scenario is divided into two phases. The first phase is the initial deployment, where UAVs are placed using partial knowledge of the disaster scenario. The second phase addresses the adaptation to changing conditions where UAVs move according to a local search algorithm to find positions that provide better coverage of victims. The suggested approach was evaluated under different conditions of scenarios. The number of UAVs have demonstrated a high degree of coverage of "victims."

From a Big Data Analytics perspective, the goal of the application is to determine optimum or near-optimum solutions in a potentially very large and complex search space. This is due to the high dimensionality and huge increase of parameters and combinatorics, with the increase in the number of UAVs and size and resolution of the disaster terrain. Therefore, this is considered an application of data analytics, namely prescriptive or decision analytics using computational intelligence techniques.

Chapter 5 proposes a novel health data analytics application based on deep learning for sleep apnea detection and quantification using statistical features of ECG signals. Sleep apnea is a serious sleep disorder phenomena that occurs when a person's breathing is interrupted during sleep. The most common diagnostic technique that is used to deal with sleep apnea is polysomnography (PSG), which is done at special sleeping labs. This technique is expensive and uncomfortable. The proposed method in this chapter has been developed for sleep apnea detection using machine learning and classification including deep learning. The simulation results obtained show that the newly proposed approach provides significant advantages compared to state-of-the-art methods, especially due to its noninvasive and low-cost nature.

Chapter 6 presents an analysis of the core concept of diagnostic models, exploring their advantages and drawbacks to enable initialization of a new pathway toward robust diagnostic models that overcome current challenges in headache disorders. The primary headache disorders are the most common complaints worldwide, and the socioeconomic and personal impact of headache disorders are very significant. The development of diagnostic models to aid in the diagnosis of primary headaches has become an interesting research topic. The chapter reviews trends in this field with a focus on the analysis of recent intelligent systems approaches with respect to the diagnosis of primary headache disorders.

This chapter demonstrates a novel Resource Allocation Scheme (RAS) and algorithm along with a new 5G network slicing technique based on classification and measuring the data traffic to satisfy QoS for smart systems such as smart healthcare application in a smart city environment. The chapter proposes the RAS for efficient utilization of the 5G radio resources for smart devices communication.

Chapter 7 reports on an application of Big Data analytics in education. The past decade witnessed a very significant rise in the use of electronic devices in education at all educational levels and stages. Although the use of computer networks is an inherent feature of online learning, the traditional schools and universities are also making extensive use of network-connected electronic devices such as mobile phones, tablets, and computers. Data mining and Big Data analytics can help educationalists to analyze enormous volume of data generated from the active usage of devices connected through a large network. In the context of education, these techniques are specifically referred to as Educational Data Mining (EDM) and Learning Analytics (LA). This chapter discusses major EDM and LA techniques used in handling big data in commercial and other activities and provides a detailed account of how these techniques are used to analyze the learning process of students, assessing their performance and providing them with detailed feedback in real time. The technologies can also assist in planning administrative strategies to provide quality services to all stakeholders of an educational institution. In order to meet these analytical requirements, researchers have developed easy-to-use data mining and visualization tools. The chapter discusses, through relevant case studies, some implementation of EDM and LA techniques in universities in different countries.

Chapter 8 attempts to address some of the challenges associated with Big Data management tools. It introduces a scalable MapReduce graph partitioning approach for high-degree vertices using master/slave partitioning. This partitioning makes Pregel-like systems in graph processing, scalable and insensitive to the effects of high-degree vertices while guaranteeing perfect balancing properties of communication and computation during all the stages of big graphs processing. A cost model and performance analysis are given to show the effectiveness and the scalability of authors' graph partitioning approach in large-scale systems.

Chapter 9 presents a multivariate and dynamic data representation model for the visualization of large amount of healthcare data, both historical and real-time for better population monitoring as well as for personalized health applications. Due to increased life expectancy and an aging population, a general view and understanding of people health are more urgently needed than before to help reducing expenditure in healthcare. The chapter proposes a multivariate and dynamic data representation model for the visualization of large amounts of healthcare data, both historical and real time.

Chapter 10 presents the adaptation of the big data analytics methods for software reliability assessment. The proposed method uses software with similar properties and known reliability indicators for the prediction of reliability of a new software. The concept of similar programs is formulated on the basis of five principles. Search results of similar programs are described. Analysis, visualization, and interpreting for offered reliability metrics of similar programs are executed. The chapter concludes with reliability similarity for comparable software based on the use of metrics for prediction of new software reliability. The reliability prediction presented in this chapter aims at allowing developers to operate resources and processes of verification and refactoring potentially increasing software reliability and cutting development cost.

Abu Dhabi, UAE   Mohammed M. Alani
Leeds, UK   Hissam Tawfik
Dubai, UAE   Mohammed Saeed
San Jose, CA, USA   Obinna Anya

Contents

Chapter 1
Big Data Environment for Smart Healthcare Applications Over 5G Mobile Network

Mohammed Dighriri, Gyu Myoung Lee, and Thar Baker

1.1 Introduction

Due to the fast growth of wireless network technologies (e.g. 5G) and ever-increasing demand for services with high quality of service (QoS) request [1], the managing of network resources becomes a permanently more challenging step that requires being correctly designed in order to advance network performance. It is also expected that the smart devices data traffic will rise quickly due to the growing use of the smart devices (e.g. smartphones, traffic control and blood pressure sensor) in numerous applications. The applications' areas of smart devices contain, for example, smart office, smart traffic monitoring, smart alerting system, smart healthcare system and logistics system [2, 3]. Furthermore, smart devices communication offers ubiquitous connectivity between smart devices that allows the interconnection of devices, for instance, laptops, smart sensors, computers, etc., to perform several automatic operations in various smart device applications. In this situation, network slicing [4] is getting an always-increasing importance as an effective approach to introducing flexibility in the management of network resources. A slice is a gathering of network resources, selected in order to satisfy the demands (e.g. in terms of QoS) of the service(s) to be delivered by the slice [5, 6]. The aim of slicing is to introduce flexibility and higher utilization of network resources by offering only the network resources necessary to fulfil the requirements of the slices enabled in the system.

An assisting aspect of network slicing is the virtualization of network resources, which allows network operators to share the common physical resources in a flexible, dynamic manner in order to utilize the existing resources in a more effective

M. Dighriri (✉) · G. M. Lee · T. Baker
Department of Computer Science, Liverpool John Moores University, Liverpool, UK
e-mail: M.H.Dighriri@2015.ljmu.ac.uk; G.M.Lee@ljmu.ac.uk; T.baker@ljmu.ac.uk

© Springer International Publishing AG, part of Springer Nature 2018
M. M. Alani et al. (eds.), *Applications of Big Data Analytics*,
https://doi.org/10.1007/978-3-319-76472-6_1

approach [7]. In our proposal, 5G radio resources are efficiently utilized as the smallest unit of a physical resource blocks (PRBs) in a relay node by allocating the data traffic of several devices as separate slices based on QoS for each application. Virtualization of network resources is presently investigated in literature particularly by concentrating on the virtualization of network functionalities [7–9]. Due to the various QoS demands and the limitation of network resources, competently allocate network resources between service slices and user equipment (UEs) are a major issue [11, 12].

1.1.1 Smart Devices

Smart devices convey small- and large-sized data with diverse QoS requirements. For instance, smart healthcare devices transmit small-sized data but are delay sensitive. The physical resource block is the smallest radio resource, which is allocated to a single device for data transmission in 4G or 5G. In the smart device applications with devices transmit small-sized data, the capacity of the PRB is not fully utilized. This results in significant degradation of the system performance. This chapter proposes a RAS for efficient utilization of the 5G radio resources for smart devices communication. In the proposed scheme, 5G radio resources are efficiently utilized by aggregating the data of several smart devices. The resources are shared by the smart devices to improve the spectral efficiency of the system.

1.1.2 Future Challenges

In mobile networks with long-term evolution (LTE) and 5G massive access such as human to human (H2H), smart devices and personal devices can lead to serious system challenges in terms of radio access network (RAN) overload and congestion. Since radio resources are an essential component and hardly exist, therefore, the efficient utilization of these radio resources is required. The novel communication technologies, such as LTE, long-term evolution advanced (LTE-A) and 5G, make use of multiple carriers schemes to offer better data rates and to ensure high QoS. The smallest resource unit allocable in the 5G system to a smart device is the PRB as illustrated in Fig. 1.1. Under favourable channel conditions, PRB is able of transmitting numerous kilobytes of data. These multiple carriers' schemes are able of transmitting a large amount of data. However, in the case of smart devices communication, both narrowband and broadband applications have to be considered to enhance QoS requirements. Especially, these applications have different size of data traffic, which need QoS specifications such as real time, accuracy and priority.

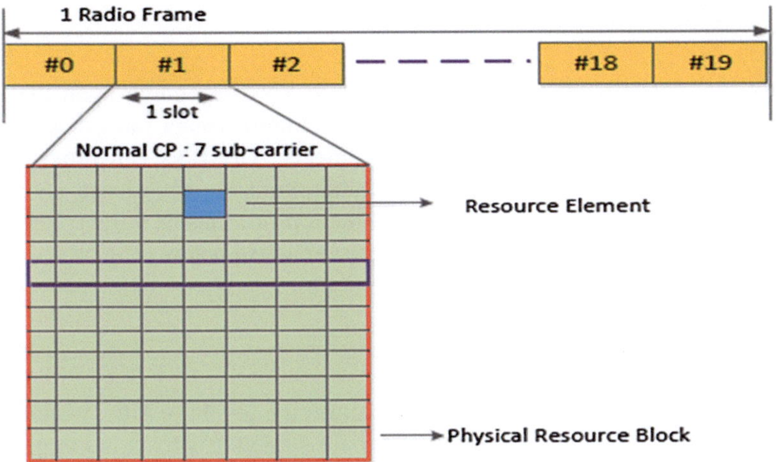

Fig. 1.1 Physical Resource Block (PRB)

If one PRB is allocated to a single smart device for data transmission of just a few bytes, then it might cause severe wastage of radio resources; also, the different types of data traffic should be considered in 5G slices approach. Therefore, the full radio resources utilization and data traffic classification should be a brilliant solution data traffic explosion and the fairness of services in the near future.

1.2 Background

1.2.1 5G Enabling Technologies

5G specified the next-generation network requirements and components in its Release 8. Those main objectives include LTE and SAE for the specification of Evolved Packet Core (EPC), Evolved UMTS Terrestrial Radio Access Network (E-UTRAN) and E-UTRA. The communication between UE and E-UTRAN is accomplished using IP, which is delivered by the EPS. In 5G, air interface and radio access networks are modified, while the architecture of EPC is kept almost the same. The EPS is the basis for LTE, LTE-A and 5G networks. The main 5G features include carrier aggregation (CA), enhanced multiple-input multiple-output (MIMO) technology, coordinated multi-point (CoMP) and relay node (RN). We will give more details about each technology in future such as CA, MIMO techniques and CoMP. Moreover, 5G will support by small cells such as Pico, Micro, Femto and RN, as we have used the RN cells for the aggregation of smart devices data traffic as describe in the following [14].

1.2.2 Infrastructure-Based RNs

The RNs are categorized into fixed and mobile RNs depending upon the infrastructure. RNs are used in distinct scenarios to improve data rates, coverage and to facilitate UEs indoor and outdoor movements. The RNs can provision UEs movements from indoor to outdoor. In addition, UEs experience satisfactory coverage through mounted RNs such as at the top of a bus or a train. The further classifications of the infrastructure-based RNs are given below [15].

1.2.2.1 Fixed Relay Nodes

Fixed RNs are mainly used to advance the coverage for those UEs, which are not close to the regular donor eNB (DeNB), or base station usually exists at the corner of the cells. Furthermore, the coverage holes due to shadowing are also improved. Fixed RNs can extend the cell coverage for the users outside the coverage of the regular base stations, as shown in Fig. 1.2, the functionalities of fixed RNs. The fixed RNs contain comparatively small antennas as compared to the antennas at the base stations. The RNs antennas are normally positioned at the top of a building, tower, poles, etc.

1.2.2.2 Mobile Relay Nodes

According to [16], 3GPP has considered mobile RNs to provide satisfactory services to the users in fast moving trains. However, in the recent literature, it has been shown that the mobile RNs can also professionally improve the services in public vehicles,

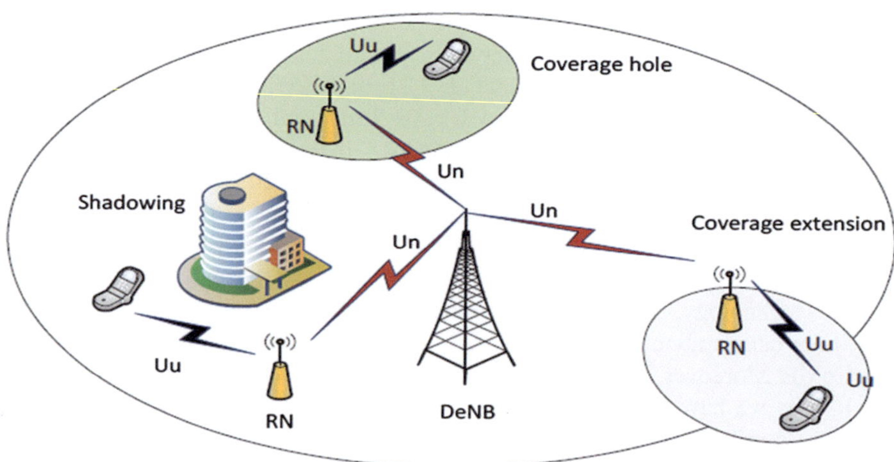

Fig. 1.2 Fixed RN

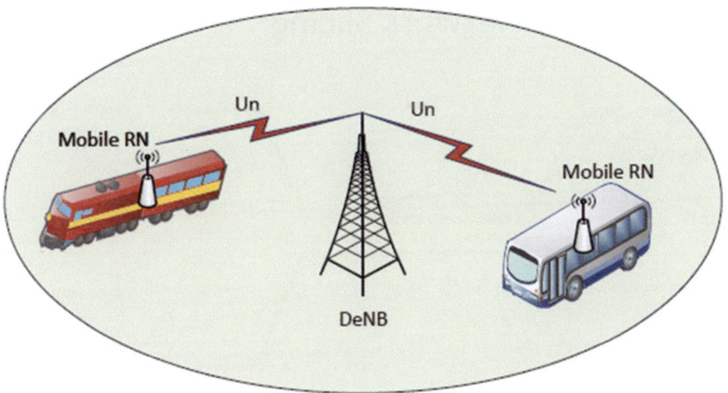

Fig. 1.3 Mobile RN

for instance, buses and trams. The purpose of mobile RNs is to offer coverage within a moving environment. The mobile RNs are positioned on the vehicle, train, etc. and create a communication path between the mobile UEs and the base station. The RNs communicate with the base station through the mobile relay link (backhaul) whereas using access link with the mobile UEs. Due to the vehicle restrictions and other safety measures, antenna size of the mobile RNs is kept small; the functionalities of mobile RNs are shown in Fig. 1.3.

1.2.3   5G Network Slicing

5G as a new generation of the mobile network is being actively discussed in the world of technology; network slicing surely is one of the most deliberated technologies nowadays. Mobile network operators such as China Mobile and SK Telecom and merchants such as Nokia and Ericsson are all knowing it as a model network architecture for the coming 5G period [17]. This novel technology allows operators slice one physical network among numerous, virtual, end-to-end (E2E) networks, each rationally isolated counting device, access, transport and core networks such as separating a hard disk drive (HDD) into C and D drives and devoted for diverse kind of services with different features and QoS requirements. Every network slice and committed resources, for example, resources within network functions virtualization (NFV), software-defined networking (SDN), cloud computing, network bandwidth, QoS and so on, are certain as seen in Fig. 1.4 [18, 19].

1.2.3.1   Data Traffic Aggregation Model

The proposed model is relying on aggregating data from several smart devices at the Packet Data Convergence Protocol (PDCP) layer of the RN. The PDCP layer

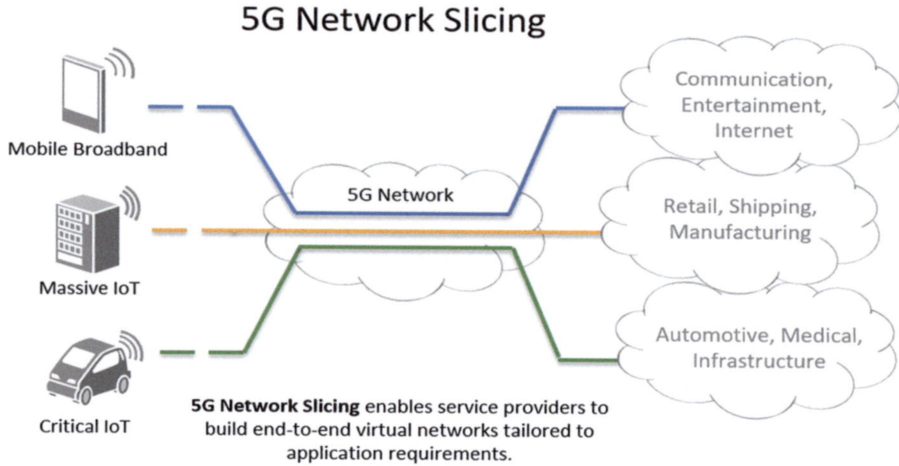

Fig. 1.4 5G network slicing

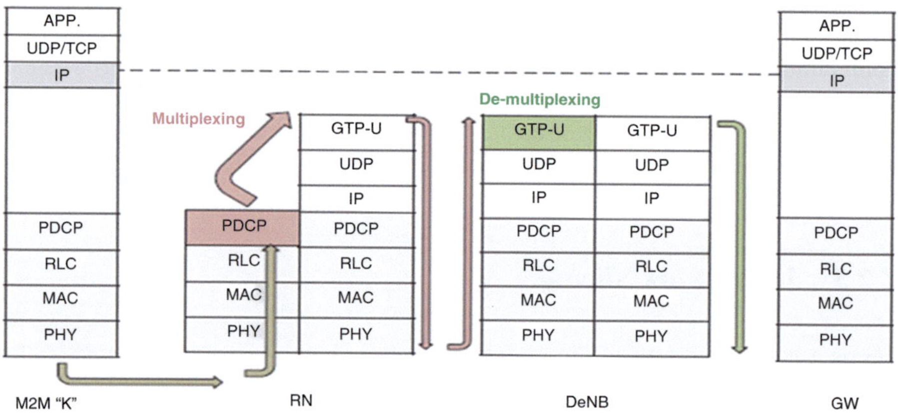

Fig. 1.5 Smart devices data packets flow diagram

performs header compression, retransmission and delivery of PDCP Session Data Units (SDUs), duplicate detection, etc. In the proposed model, PDCP layer is used for the aggregation of the smart devices data in the uplink. The main reason for selecting PDCP for aggregation in the uplink is to aggregate data with a minimum number of the additional headers as shown in Fig. 1.5.

The individual data packets from the several smart devices approach the PHY layer of aggregation device with various intact headers such as Medium Access Control (MAC), Radio Link Control (RLC) and PDCP. The headers are removed as the received data is transported to the upper layers. Upon of the data packets arrival toward PDCP, all the headers are removed, and only the payload from the individual devices are available, which are aggregated.

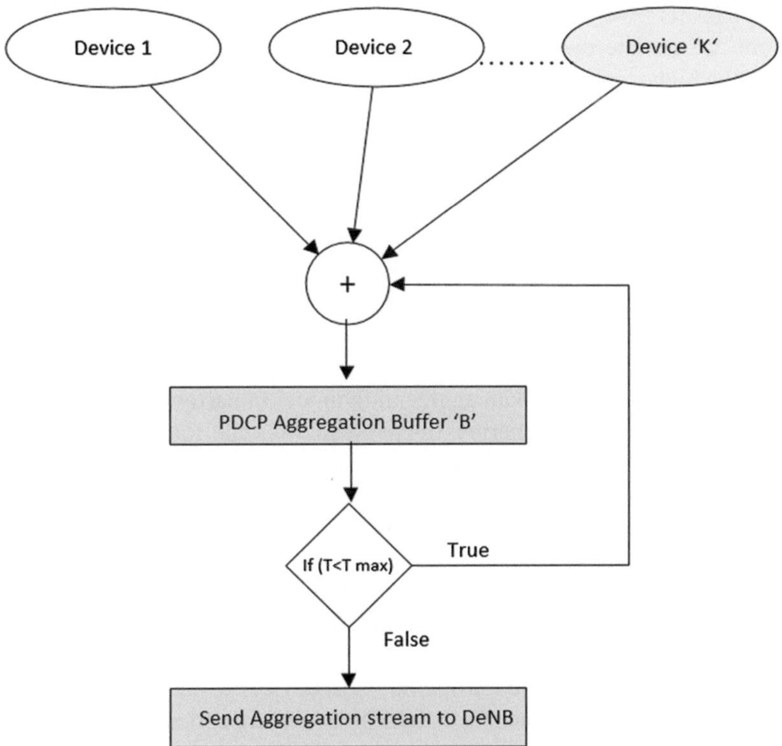

Fig. 1.6 Smart devices data aggregation algorithm

One single aggregation buffer *B* at the RN is considered to aggregate smart devices data traffic. This buffer aggregates data from different smart devices ensuring QoS for both the 5G and smart devices data traffic. In this implementation, RN is used for smart devices and base station for 5G data traffic. In order to reach the maximum performance improvements in spectral efficiency, packet propagation delay and cell throughput, we consider scenarios in which all the smart devices communicate with the base station through a RN. The smart devices data aggregation algorithm is shown in Fig. 1.6 and described as follows:

- Data from *K* smart devices are considered for aggregation.
- The essential parameter for smart devices data aggregation is the maximum delay time *Tmax* for the packet at the RN.

The maximum delay time *Tmax* is an essential parameter for smart devices data and is calculated according to the various traffic classes of the smart devices. Smart devices data have different priorities according to their applications. For example, data packets received from the smart devices deployed in smart healthcare system scenario for the measurement of temperature or pulse rate of the patient have high priority over the packets from smart devices, which are deployed in smartphones.

The data packets from a device having the highest priority face the smallest delay. Therefore, we initiate the *Tmax* value as the inter-send time of the smart devices data with the highest priority. For example, in the simulation setup for distinct smart device applications, the inter-send time of the smart devices traffic model is 1 s, which is the maximum time a packet is delayed at the RN. Thus, the value of the *Tmax* is initiated as 1 s, which means that the data packets received from the distinct smart devices are delayed for 1 s at the RN.

The value of *Tmax* is adaptive, i.e. the algorithm updates the value of *Tmax* if RN receives packets from a device, which has higher priority than the priorities of all the other devices in the queue of the RN. The data from all the smart devices are buffered at the RN. The individual IP headers of all the smart devices are kept intact. The data packets are buffered until time delay approaches *Tmax*. In order to compare the performance of data aggregation model in narrowband and broadband smart devices application scenarios, the aggregation scale for smart device is kept 1 (unaggregated), 5, 10, 15 and 20 in both cases. The aggregation scale represents the number of devices, which are aggregated. For example, in a scenario with 180 smart devices, the aggregation scale of 5, 10, 15 and 20 means that the data from the group of 5, 10, 15 and 20 devices is aggregated at the RN, respectively.

The aggregated data is sent to the base station through the *Un* interface where the data is de-multiplexed. The individual IP streams are then sent to the respective application server by the base station.

The smart device packets flow from the smart devices to the aGW through RN. K smart device transmits data packets to the RN, which are collected at the PHY layer of the RN. The packets are transported to the PDCP layer of the RN on the uplink. The IP packets are packed according to their quality control identifier (QCI) values in the aggregation buffer. The aggregation buffer collects packets from several smart devices. The data packets are placed in the aggregation buffer according to the packet arrival from the different devices. The detailed structure of the aggregated data Model is depicted in Fig. 1.5 ,where only the layer two protocols are presented to illustrate the aggregation of the smart devices data. The RN PHY layer receives the data packets in the form of distinct transport block size (TBS). The TBS is shown from 1 to K, which shows the TBS transmitted by the smart devices at the RN. The data packets arrive at the RLC through MAC layer. The RLC headers are removed, and the remaining protocol data unit (PDU) is transported to the PDCP. The received PDUs at the PDCP layer comprised of the individual IP headers of each smart devices and pack into single PDCP buffer.

1.2.4 Resource Allocation Scheme (RAS)

The application layers in the 5G mobile networks are the main terminal to offer exceptional QoS over different and variety of networks for smart devices. The proposed RAS will be based on data traffic aggregation and multiplexing models as we mentioned above, which is focused on service layers, based on QoS requirements for each service (application) layer. Therefore, we will clarify the main 5G network architecture layers, which are physical/MAC layers, network layers, open transport protocol (OTA) layers and service layers.

In this case, more study is needed on the virtualization of radio resources in order to perform the resource allocation scheme (RAS) for network slices. Certainly, the main aspect to be considered is the way radio resources are allocated to dissimilar slices in order to achieve the requirements of such slices. The duty relevant to (RAS) becomes more challenging with network slicing, as it introduces a two-tier priority in the system. The first tier refers to the priority of different slices, i.e. inter-slice priority, as each slice has its own priority defined according to the agreements between the network provider and the slice owner. The second tier refers to the priority between the users of the same slice, i.e. intra-slice priority. Once looking at the solutions exploited over existing 4G systems to cope with radio resources, it obviously emerges that 4G networks are able to maximize the QoS of the served users and, however, are not capable of performing the resource allocation in slicing environments [13]. This limitation is due to the fact that RAS in 4G systems is performed by assigning the priorities to the requested services via the UE. This method thus fails when considering that in 5G systems different UEs may belong to different slices with different priorities, and thus such UEs should be managed by considering the priority of the slice they belong to plus the priority of the service they need.

In this chapter, we propose a novel RAS; as shown in Fig. 1.7, it exploits a two-tier priority levels. Our proposal relies on the idea that network slices communicate to an admission control entity with the desired QoS level. The RAS, based on the priority of the slice, decides about serving the slice. Finally, according to the inter-

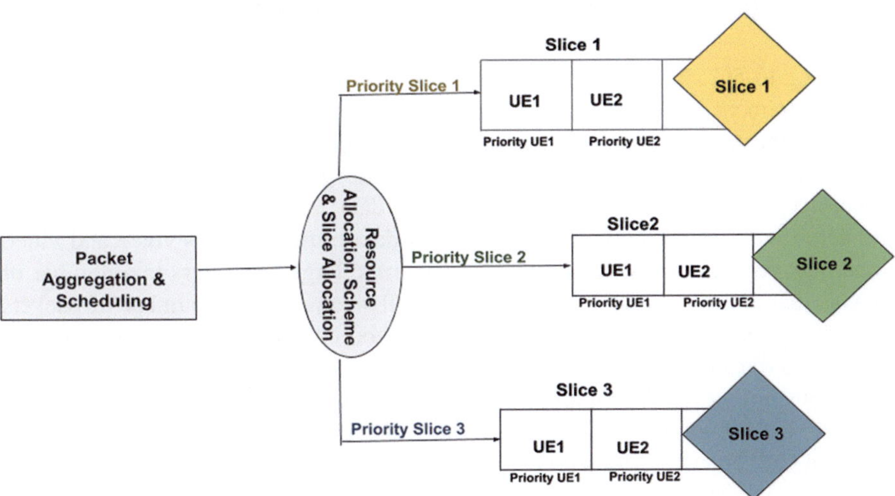

Fig. 1.7 RAS with inter-slice and intra-slice priority

and intra-slice priority, the virtual network allocates the physical radio resources to the UEs of the admitted slices. According to the decision of the RAS, the resource allocation mission is performed with the purpose to maximize the quality of experience (QoE) of the users inside each slice, by considering the inter-slice priority. In this chapter, the QoE is measured by considering the effective throughput experienced by the users, normalized according to their maximum demanded data rate. With this target, the resources allocated to a slice with low priority could be reduced, if needed, down to the minimum amount capable of meeting the basic QoS requirements to admit new slice(s) with higher priority. Therefore, doing our proposal dynamically changes a number of network resources allocated to network slices.

According to the packets load without affecting the QoE of the users and while improving the network utilization. To summarize, the main contributions of this chapter could be listed as follows:

- A novel RAS with two-tier priority level has been proposed in our virtualized 5G system model.
- The proposed RAS dynamically sets the resources allocated to allow slices according to the current traffic load and based on efficiently utilizing the smallest untie of PRB by aggregating the data of several devices.
- Inter-slice and intra-slice priority order have been considered into account for assigning the QoE maximization problem of resource allocation task. Since priority orders for QoE purpose can advance the satisfactory level of UEs and network utilization.

1.3   Resource Allocation Scheme Environment

According to 5G slicing technology, we will focus on classifying and measuring QoS requirement and data traffic of smart device applications such as smartphones, smart healthcare system and smart traffic monitoring (Fig. 1.8). As results of smart device data traffic characteristics in 5G network slicing framework, such as the content type of data, amounts typed of flow data, priority of data transmission and data transmission mode. Content type of data traffic contains voice and video streaming; amount type consists of different sizes: large size refers to a number of packets that are more than 1 K bytes and small size refers to a number of packets that are less than 1 K bytes. Transmission method contains periodic transmission, continuous transmission, burst transmission and time-response transmission; priority of transmitting consists of low, medium and high. Depending on the smart device applications, slicing our research would have classified them into three main slices based on QoS and data traffic types.

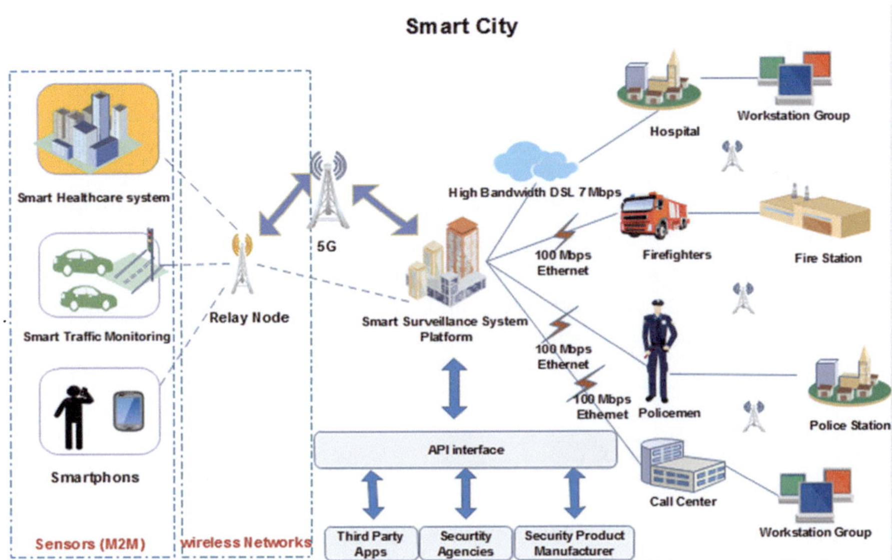

Fig. 1.8 Smart systems in smart city use cas*e*

1.3.1   Related Works

In a literature review, numerous solutions for efficiently enhancing virtualization of network resources have been considered to improve the QoE of UEs and network resource utilization [9]. A competent wireless network virtualization for LTE systems has been suggested in [10], which proposes a slicing structure to efficiently allocate physical resource blocks to diverse service providers (SPs) in order to maximize the utilization of resources. The approach is dynamic and flexible for addressing arbitrary fairness requirements of different SPs. Correspondingly, [20] proposed a framework for wireless resource virtualization in LTE system to allow allocation of radio resources among mobile network operators. An iterative algorithm has been proposed to solve the Binary Integer Programming (BIP) with less computational overhead. However, above considered schemes do not take the priority among different slices, besides the priority among the users within the same slice.

For the limitation of network resources, the RAS can be executed to improve communication reliability and network utilization. In [21], a combined resource provisioning and RAS have been proposed targeting to maximize the total rate of virtualized networks based on their channel state information. An iterative slice provisioning algorithm has been proposed to adjust minimum slice requirements based on channel state information but without considering global resource utilization of the network as well as inter- and intra-slice priority.

In [21], a scheme for allocating downlink network resources has been proposed. The scheme decides to accept a novel service only if the provisioning of this new service does not affect the throughput of the services in the cell. Consequently, this work does not take into consideration the dynamic modification of the QoE experienced by mobile users in order to increase network capacity and resource utilization.

Centralized joint power and RAS for prioritized multi-tier cellular networks have been proposed in [21]. The scheme has been developed to admit users with higher-priority requirement to maximize the number of users. In this case, the priority is only considered at the user level, and, thus, this work fails in guaranteeing differentiation in case users belong to slices with different priorities.

1.3.2 System Models

As depicted in Fig. 1.9, our model consists of four main elements: the service slice layer, the virtual network layer, the physical resources and the RAS.

1.3.2.1 Service Slices

The service slices offer different services (e.g. smartphones, smart traffic monitoring and smart healthcare system) which need resources to be served. We designate with $\mathbb{S} = \{1, 2, 3... S\}$ the set of slices in the virtual network. Each slice s has a set of UEs, such a set is symbolized by $\mathbb{U}s = \{1, 2... Us\}$. Each slice s performs a request to the RAS in terms of QoS restraints. In this chapter, we model such a request with

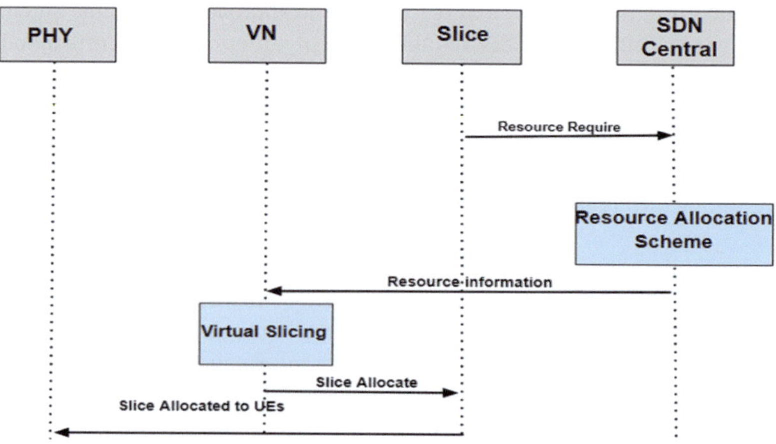

Fig. 1.9 Flow of RAS

R_Smins and R_Smax, which denote the minimum and maximum data rates associated with the slice s, respectively. Each slice s is characterized by a priority, ιs, where such priorities are defined with the constraint that $\sum_{s \in S} \mathbb{p}s = 1$. Similarly, each user u belonging to the slice s, i.e. us, is characterized by a priority μus, where $\sum_{us \in Us} \mu us = 1$.

1.3.2.2   Virtual Network

The virtual network layer delivers an abstraction of the physical network resources. According to the decisions of the admission control, the virtual network slices the resources of the network to accommodate different slices. The virtual network receives the requests of different slices in terms of UEs to be served for each slice and executes the subsequent allocation of physical resources according to the inter- and intra-slice priority while considering the QoE of UEs.

With this aim, (1.10), we can define:

$$qus = \left(\frac{rus}{R_S \max} \right) \tag{1.1}$$

As the QoE of UE u in the slice s; rus is the data rate of the UE u in the slice s. The overall s; QoE us is the data rate of the of users, belonging to slice s can be computed as:

$$qs = \sum_{us \in Us} (qus)\mu us \tag{1.2}$$

Finally, we can define:

$$Q = \sum_{s \in S} (qs)\rho s \tag{1.3}$$

as the general QoE experienced by all the UEs of all slices. The virtual network assigns the resources on a scheduling frame basis. We outline with, q^t_{us}, q^t_s and Qt the QoE in a generic scheduling frame t. Accordingly, we can also define the time-average QoE values as follows:

$$E\,[qus] = \frac{1}{T}\,q^t_{us} \tag{1.4}$$

$$E\,[qs] = \frac{1}{T}\,q^t_s \tag{1.5}$$

$$E\ [Q] = \frac{1}{T}\ Qt \qquad (1.6)$$

where T is the overall number of considered scheduling frames.

1.3.2.3 Physical Resources

The physical resources denote the radio resources available in the virtual network. For the purpose of simplicity, we refer to the downlink channel of one macro-cell. The total available bandwidth is indicated by B MHz. The set $\mathbb{M} = \{1, 2...M\}$ represents the available subchannels, where the bandwidth of the generic subchannel m is $bm = \frac{B}{M}$. The total transmit power $PTOT$ is uniformly allocated to each subchannel, i.e. $pm = \frac{B}{M}$.

When PM is assigning the physical resources, we consider the channel conditions of the UEs. We assume that channel condition is determined by transmission path loss and shadowing components [22]. The path loss and the shadowing fading path loss are assumed to be a Gaussian random variable with zero mean and σ standard deviation equal to $8dB$ [22]. So, the path loss is based on the distance value dus between a generic UE and the macro-cell, which is given in Eq. 1.7.

$$PL(dus) = 128.1 + 37.6\log 10(dus) + \log 10(Xus) \qquad (1.7)$$

where UE Xus is the log-normal shadow fading path loss of UE [22]. We also assume that the macro-cell receives perfect channel gain information from all UEs belong to different service slices, where hm, us is the subchannel gain for the UE u within slice s and can be defined as $hm, us = 10 - PL(dus)/10$ [22]. The data rate of the UE with a slice s, denoted with rus, can be defined in Eq. 1.8 [23].

$$rus = \sum_{m \in M} \alpha m, us bm \left(1 + \frac{pm \mid hm, us \mid 2}{N0bm}\right) \qquad (1.8)$$

where $N0$ is the noise spectral density and $\alpha m, us$ is the situation of the UE us which has been described in Eq. 1.9.

$$\alpha m, us = \begin{cases} 1 \\ 0 \end{cases} \text{ if sub} - \text{channel } m \text{ is assigned to } us \text{ otherwise} \qquad (1.9)$$

1.3.3 Two-Tier Scheme and Resource Allocation

In this section, we describe our proposed approach for two-tier admission control and resource allocation based on services allocation.

1.3.3.1   Services Allocation

The 5G mobile network terminal offers exceptional QoS through a diversity of networks. Nowadays, the mobile Internet users choose manually the wireless port of different Internet service providers (ISP) without having the opportunity to exploit the QoS history to choose the suitable mobile network linking for a provided service. In the future, the 5G phones will offer a chance for QoS analysis and storage of measured data traffic in the mobile network terminal. There are diverse QoS parameters (e.g. bandwidth, delay, jitter and reliability), which will support in future of 5G mobile running in the mobile terminal. System processes will offer the best appropriate wireless connection based on needed QoS automatically. Therefore, we will consider various types of priorities as service allocation as shown in Figs. 1.10 and 1.11 [23]. These priority types based on different QoS requirement by various users and services.

Smartphones

Smartphones and tablets are recent technologies that are represented as popular data traffic. Although smartphones are expected to continue as the key personal device and have more development in terms of performance and ability, the number of personal devices growth was driven by such devices as wearable or sensors to reach millions in 2020. In these devices, the content type of mobile streaming is video; the total of the flow packets is regularly numerous megabytes or even tens of megabytes; it is many of packets; the transmission way is usually continual transmission; the priority is generally low due to the video requires broad bandwidth and is likely to be blocked in congestion [1].

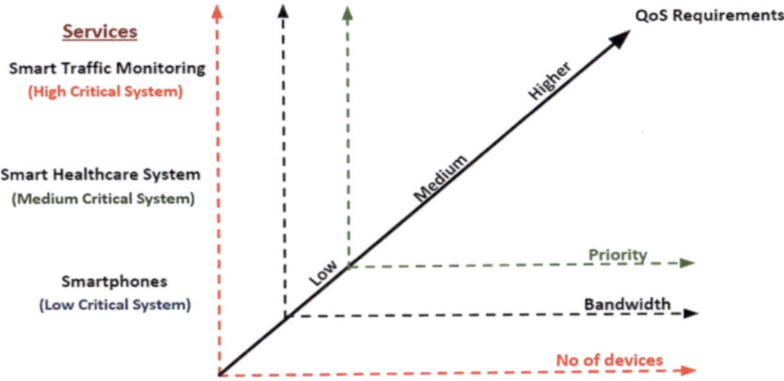

Fig. 1.10 Services allocation priorities

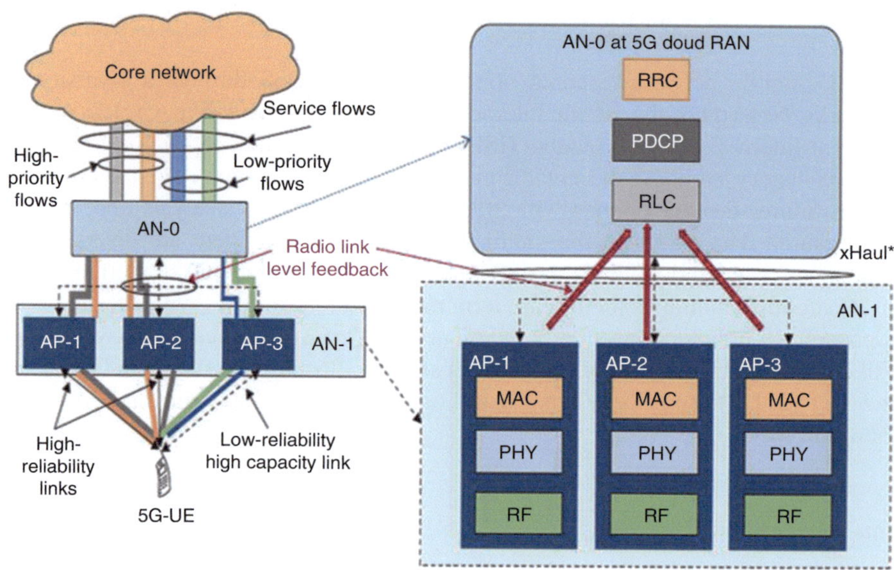

Fig. 1.11 Services allocation priorities architectural review

Smart Healthcare System

The smart healthcare system as sensitive data traffic is a promising model, which has currently achieved extensive attention in research and industry. A sensor body area network (BAN) is generally positioned nearby the patient to gather information about the numerous health parameters, for instance, blood pressure, pulse rate and temperature. Moreover, the patients are also monitored repeatedly by placing smart device sensors on the body of the patient when they are outside the hospitals or home. For handling critical situations, alarms are triggered to send messages to the related physicians for urgent treatment [4]. In a smart healthcare system scenario, in order to monitor the patients frequently outside the medical centres (e.g. hospitals), the patients are equipped with smart devices that monitor various health parameters.

Smart Traffic Monitoring

Smart traffic monitoring allows the conversation of alerted information between vehicles infrastructure and the system applications over communication approaches and technologies. In this system, we will consider heavy data traffic. Vehicles connect with other vehicles (V2V) or communicate with smart traffic monitoring servers, vehicle to infrastructure (V2I). This system application includes the collision prevention and safety, parking time, the Internet connectivity, transportation time, fuel consumption, video monitoring, etc. [1]. In the case of emergency, the information from devices positioned to monitor emergency situations is transmitted to other networked vehicles within the communication range. To prevent any more accidents, the connection between the vehicles and the servers should be very fast for the detection of emergency messages and delivery of alerting messages. Since the reply time of the warning messages is very small, collision avoidance services request a high level of QoS (i.e. low latency), which can be supported by the 5G cellular networks. According to [1], the alerting messages are small size and must only be sent in serious circumstances for effective using of the communication network bandwidth. Traffic and infrastructure management play an important role in monitoring the issue of traffic congestion.

1.3.3.2 Service Slices Strategy

A RAS based on priority has been designed in algorithm (Table 1.1.) This scheme can be used to cope with the entrances of new slices or users and provides a global optimization of the resources allocated to service slices. For the purpose of simplicity, algorithm 1 denotes to the RAS of novel UEs belonging to the same slice. The steps of our proposed RAS can be applied for admission control of new slices, by simply adjusting the parameters under consideration. When the new UE arrives at the network, by considering the QoE of the users in the same slice, we can derive an acceptance probability of the novel user in the virtual network by considering the constraints in terms of intra-slice priority as well as the QoE of served UEs. In our RAS, new UEs are accepted if the existing resources are sufficient to guarantee to satisfy at least the demand on the minimum data rate. The set of accepted users is thus offered as input to the resource allocation process.

1.3.3.3 Resource Allocation

The overall problem under consideration during the resource allocation step is the maximization of the QoE of UEs, by simultaneously considering the inter- and intra-slice priority. This problem can be formulated as in Eq. 1.10.

$$\text{P1}:$$

$$\text{maximize} \sum_{s \in \mathbb{S}} \left[\sum_{us \in \mathbb{U}s} \left(\frac{\nabla \cong s}{R_S \max} \right) \mathfrak{U}us \right] \rho s \qquad (1.10)$$

Subject to, (11)

$$\sum_{m \in M} \sum_{s \in S} \sum_{us \in Us} \alpha m, usbm \leq B, \qquad (1.11a)$$

Table 1.1 Resource allocation scheme (RAS)

Algorithm 1: : RAS Algorithm of New Users

```
for t := 1 to T do
    for s := 1 to S do
        for u := 1 to U do
            for m := 1 to M do
                Calculate qus∀Us ∈ Us;
                find UE xs with the max QoE;
                find UE js with the max QoE;
                while a new UE us ∈ Us enters
                the network do
                    Calculate the new QoE value of us:qus;
                    Then, find the neighbour QoE value of us:qus;
                    if qus − qus >0 then
                        if E[qus] < qus then
                            Inject UE us;
                            check priority order;
                            if the priority order are the
                            same then
                                xs will be replaced by the
                                new UE; else
                                    js will be replaced by
                                    the new UE;
                                end
                        end
                        else
                            Do not admit UE us;
                        end
                    end
                    else
                        generate accept probability
                        p = −ΔQoE / T ;
                        then, the new UE will be
                        rejected based on the
                        probability p;
                    end
                end
            end
        end
    end
end
```

$$R_S \min \leq rus \leq R_S \max, \tag{1.11b}$$

where constraint (1.11a) indicates that a number of allocated subchannels cannot overcome the maximum available bandwidth; this constraint implicitly refers to the orthogonally of assigned resources, too. Constraint (1.11b) indicates that the received the associated data rate by UE us is restricted by the requirements of the associated slice s. It is useless that, in Eq. 1.10, the QoE is a number lower or equal than 1; as a consequence, the higher the priority of a slice, the lower the value of ρs. This happens similarly for the users, i.e. the higher the priority of a user; the lower is the value $\mathcal{U}us$. The resource allocation procedure is performed by considering the physical resources available in the network as well as the channel conditions of the UEs.

1.4   Simulation Approach

The Optimized Network Engineering Tool (OPNET) is simulation used to assess the performance of the proposed scheme. Several scenarios are simulated to evaluate the impact of smart devices data traffic on regular 4G and 5G mobile networks data traffic. The simulated 4G and 5G data traffic classes include File Transfer Protocol (FTP), Voice over IP (VoIP) and video users. The scenarios are categorized into first scenario aggregation PRBs with RAS, second scenario aggregation PRBs without RAS and third scenario without both aggregation PRBs and RAS. The results show the significant impact of smart devices data traffic on low-priority data traffic. The end-to-end network performance has been improved by allocated data of several smart devices, which is determined by simulating several scenarios. Considerable performance improvement is achieved in terms of average cell throughput, FTP average upload response time, FTP average packet end-to-end delay and radio resource utilization [24].

1.4.1   Simulation Setup

The LTE-A node protocols, which we have developed to work with the 5G mobile network. The remote server supports email, VoIP, FTP and video applications in the form of smart systems. The remote server and the Access Gateway (aGW) are interconnected with an Ethernet link with an average delay of 20 ms. The aGW node protocols include Internet Protocol (IP) and Ethernet. The aGW and Enb nodes (eNB1, eNB2 ...) communicate through IP edge cloud (1, 2, 3 and 4). QoS parameters at the transport network (TN) guarantees QoS parameterization and traffic difference. The user mobility in a cell is matched by the mobility model by updating the location of the user at every sampling interval. The user mobility information is stored on the global server (global UE server). The channel model parameters for the air interface contain path loss, slow-fading and fast-fading models. The simulation modelling mostly focuses on the user plane to perform end-to-end performance evaluations. An inclusive explanation of the LTE-A simulation model and details about the protocol stacks can be found in [24].

The different traffics QoS have been set according to the 3GPP standardization. The other simulation parameters are recorded in Table 1.2.

Table 1.2 Simulation parameters

Parameter	Setting
Simulation length	600 s
Cell layout	1 Enb
eNB coverage radius	350 m
Min. eNB-UEs	35 m
Max. terminal power	23 dBm
5G parameters	
5G cell	8*8 antennas
Cloud	Edge cloud
Capability	Enabled
RN parameters	
PRBs for RN	3 PRBs are allocated to RN by DeNB to evaluate PRB utilization
Type of RN	Fixed
RN 1	Support by 4 antennas, 10 MHz TDD
RN2	Support by 3 antennas, 5 MHz TDD
RN3	Support by 2 antennas, 3 MHz TDD
TBS capacity	1608 bits against MCS 16 and PRBs 5
	Available service rate *TBS—overhead* (bits/TTI)
	1608 *(TBS)* − 352 *(overhead)* = *1256* bits/TTI
Simulated scenarios	Aggregation with RAS
	Aggregation without RAS
	Without Aggregation and RAS
Terminal speed	120 km/h
Mobility model	Random Way Point (RWP)
Frequency reuse factor	1
System bandwidth	5 MHz
Path loss	$128.1 + 37.6log\ 10(R)$. R in fan
Slow fading	Log-normal shadowing, correlation 1, deviation 8 Db
Fast fading	Jakes-like method
UE buffer size	∞
RN PDCP buffer size	∞
Power control	Fractional PC, $\alpha. = 0.6$, $Po = -58$ dBm
Applications	Email, VoIP, Video and FTP

1.4.2 QoS of Radio Bearers

The LTE QoS has gained considerable importance in the designing and planning of the networks. There are possibilities to use the LTE network for various operations. For example, some subscriber uses the network services for emergency cases, while others use the services for entertainment purposes. QoS explains how a network serves the subscribers due to the enclosed network architecture and protocols. In LTE, the term bearer can be defined as the flow of an IP packet between the UE and P-GW. Each bearer is linked with particular QoS parameter. The network provides almost same services to the packets which are linked to individual or same bearer. For establishing a communication path between UE and PDN, UE attempt to generate a bearer by default. Such bearers are called default bearers. The other bearers are named as dedicated bearers which are established to the PDNs. Establishing more than one bearer is possible. This is because one user demands several services, and each service demands specific bearer. For example, if a bearer is established, it is possible to generate more bearers in the presence of an existing bearer.

Moreover, the QoS value of an existing and newly created bearer is possible to vary. The bearer can be classified into Guaranteed Bit Rate (GBT) and Non-Guaranteed Bit Rate (Non-GBR).

- The GBR bearer has a minimum bandwidth which is allocated by the network for various services such as voice and video communication, regardless of that are used or not. Due to dedicated system bandwidth, the GBR bearer does not undergo any packet loss due to congestion and are free from latency.
- Non-GBR bearer is not allocated a specified bandwidth by the network. These bearers are used for best-effort services such as web browsing, email, etc. These bearers might undergo packet loss due to congestion.
- Quality control identifier (QCI) describes how the network treats the received IP packets. The QCI value is differentiated according to the priority of the bearer, bearer delay budget and bearer packet loss rate. 3GPP has defined several QCI values in LTE which are summarized in Table 1.3.

1.4.3 Radio Resource Allocation Algorithm

Packet scheduling is the distribution of radio resources between the radio bearers in a cell by the eNB. In 3GPP LTE standards, this task is performed by the MAC scheduler in the eNB. The allocation of the downlink and uplink radio resources by

Table 1.3 LTE QCI values [6]

QCI	Resource	Delay	Priority	Error	Service type
1	GBR Non GBR	100 ms	2	10−2	Conversational (VoIP)
2		150 ms	4	10−3	Conversational (Video)
3		50 ms	3	10−3	Real time gaming
4		300 ms	5	10−6	Non conversational voice
5		100 ms	1	10−6	IMS signalling
6		300 ms	6	10−6	Video Buffered streaming
7		100 ms	7	10−3	TCP based (email. HTTP, FTP)
8		300 ms	8	10−6	Voice, video and interactive gaming
9		300 ms	9	10−6	video buffering streaming

the eNB to the UEs depends upon the data present in the buffers of the eNB and the UEs, respectively. If the data for a particular UE is present in the buffer of the eNB, then the eNB allocates radio resources to the UE for downlink transmission if eNB has enough available radio resources, and the QoS requirements of the other UEs located in the coverage area of the eNB are fulfilled. Similarly, in uplink transmission, the UEs transmit Buffer Status Report (BSR) information to the eNB for granting radio resources if there is data present in the buffer of the UEs. UE BSR information also identifies the types of traffic in the UE buffer. The eNB allocates radio resources for downlink and uplink according to the radio bearers QoS requirements of the UE. Time Domain-Maximum Throughput (TD-MT) scheduler provides the radio resources to the UEs close to eNB and bears good channel conditions. The users at the cell-edge may not get radio resources. The TD-MT scheduler provides maximum throughput at the cost of fairness [25], which can be expressed simply as in Eq. 1.12:

$$PkT\ D = rk\ (t) \tag{1.12}$$

1.5 Simulation Scenarios

The performance of the proposed models will be evaluated by three scenarios relay on RNs and 5G cell. In the first scenario, an aggregation PRBs with RAS, in the second scenario an aggregation PRBs without RAS and third scenario is without both aggregation PRBs and RAS as showed Table 1.4. The data packets from all the active smart devices, which are positioned in the nearness of the RN and 5G cell, are aggregated at the RN before being sent to the DeNB. Though, only the periodic per-hop control model is used in which the large aggregated data packets are served to guarantee full utilization of PRBs. The expiry timer is presented in order to limit the multiplexing delay particularly in the low-loaded scenarios between RN and DeNB. In this situation, the aggregated packet is served after Tmax at the latest. All the overhead stated scenarios are further sub-categorized into numerous sub-scenarios. In the first sub-scenario, smart traffic monitoring devices are placed in the nearness of the RN1, which are supported by four antennas and ten MHz TDD with a low level of priority 5 ms. The second sub-scenarios smart healthcare system devices are placed in the nearness of the RN2, which are supported by three antennas and five MHz TDD with a medium level of priority 10 ms. The third sub-scenarios smartphones devices are placed in the nearness of the RN3, which are supported by two antennas and three MHz TDD with a medium level of priority 15 ms. (Table 1.4).

Table 1.4 Simulation scenarios

Scenarios	(1) Aggregation PRBs with RAS	(2) Aggregation PRBs without RAS	(3) Without both aggregation PRBs and RAS
Smart systems	All	All	All
Application types	Email, VoIP, FTP and video	Email, VoIP, FTP and video	Email, VoIP, FTP and video

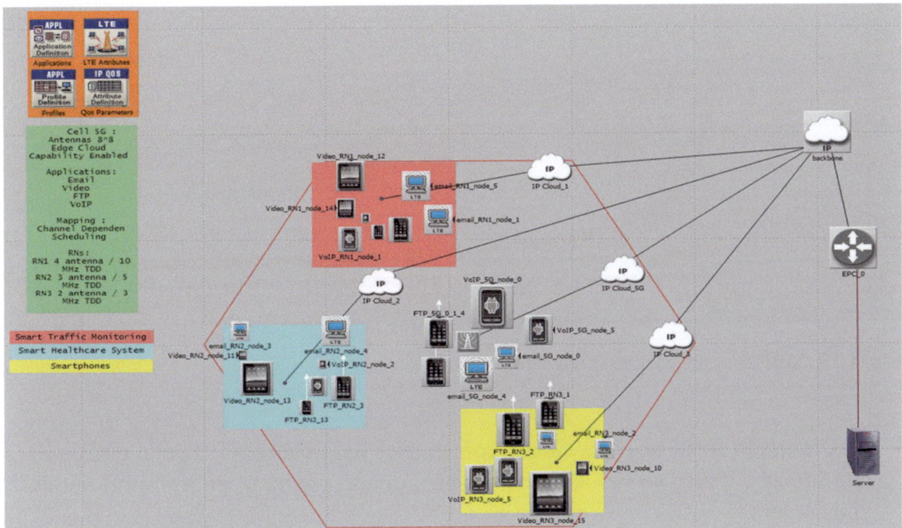

Fig. 1.12 OPNET 5G project

1.5.1 OPNET 5G Model Description

In OPNET simulation there is a scenario for LTE-A project editor with some of the most important entities of the simulation model. Whereas, the node's model of the DeNB and RNs implementation has been modified to 5G mobile network requirements, such as a number of antennas, edge cloud, small cells and high level of bandwidth as Fig. 1.12 depictsthat more description of these entities is given below:

- *Applications:* Different applications such as VoIP, video, FTP and email are defined and configured in the applications.
- *Profile:* Various traffic models are defined in profiles. Moreover, the other operating parameters such as simulation length, start time, etc. are also defined in profiles to support applications requirement.
- *Mobility:* Mobility models of various users are defined. Moreover, channel conditions such as pathloss, fading, etc. are also defined in mobility.
- *Global UE server:* Contains user's data and transport functionalities.
- *Remote server:* It is the application server.

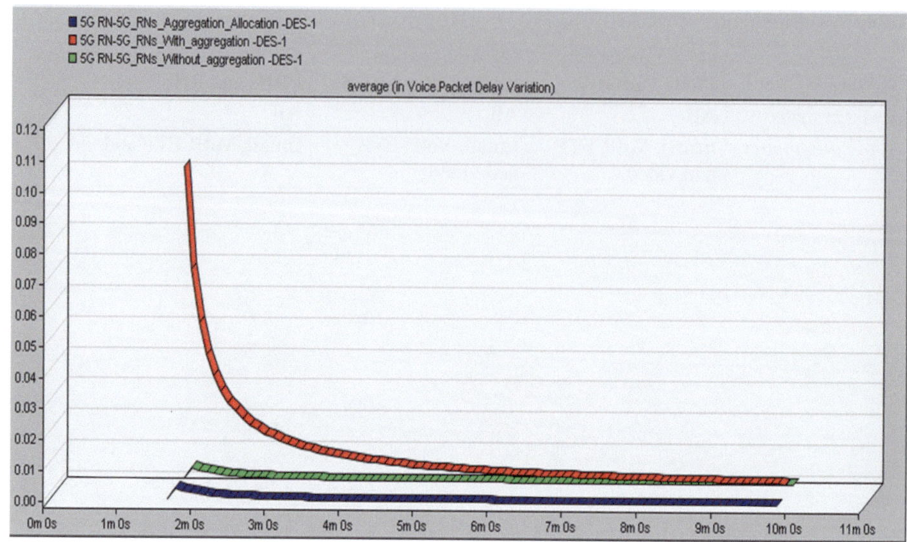

Fig. 1.13 VoIP average packets delay variation (s)

- *IP Cloud:* In form of edge clouds routes user data packets between eNBs, RNs and servers. It also serves as a peer-to-peer connector between transport network and servers.
- *Ethernet connectors (E1, E2, E3 and E4):* Are connectors in the linked network.
- *eNB:* eNB models the functionalities of eNB in E-UTRAN.
- *UE:* UEs represents different users in with various applications.

1.5.2 Experimental Results

The average air interface packet for VoIP users are shown in Fig. 1.13. The results display that the VoIP users have the diverse packets delay variation in all three scenarios even when allocated together with GBR bearers. The cause is the proportional varieties distinguishing of priority, which is characterized by RAS algorithm in "Sc1". Meanwhile, the VoIP bearer has a relatively low level of packets delay accrued data rate; it tends to get higher priority feature and will permanently be scheduled first.

The VoIP average end-to-end delay is shown in Fig. 1.14. It can be seen that "Sc1" and "Sc3" scenarios have somewhat better end-to-end delay compared to "Sc2" scenario; this is because of the fact that the "Sc1" allocate the VoIP bearers to a higher MAC QoS class by allocating this PRBs to VoIP users in this scenario.

As shown in Fig. 1.15, the average packets delay variation for the video bearers, the result describes that the video bearers have worse performance in the "Sc2" scenario compared to "Sc1" or "Sc3" scenarios where the video bearers are allocated

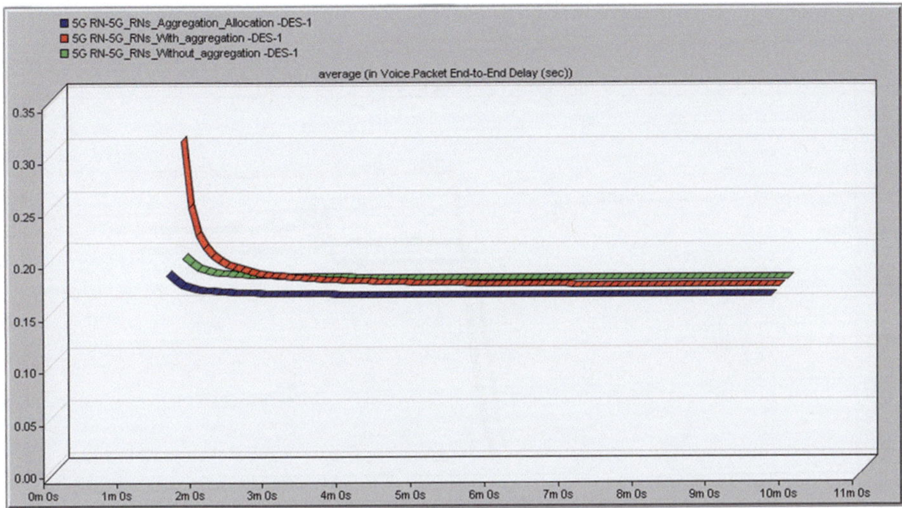

Fig. 1.14 VoIP average Packets End-to-End Delay (s)

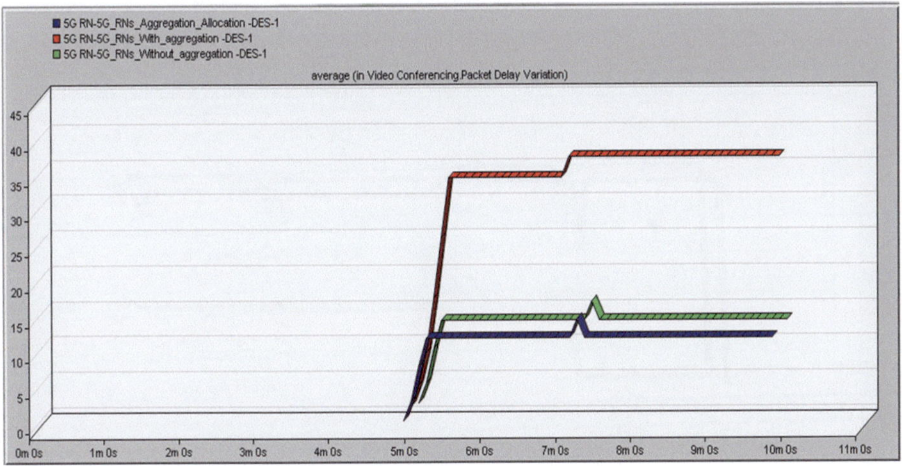

Fig. 1.15 Video average packets delay variation (s)

into the GBR MAC classes. In the "Sc3" the video bearers share the same non-GBR MAC QoS class with email, FTP and VoIP bearers since the accumulated data rate of the video. Bearers are expressively high ($\sim$ 350 kbps); they do not become served all the time.

The performance dropped down of the video bearers in "Sc2" scenario as shown obviously in Fig. 1.16 with the average end-to-end delay. The video bearers suffer from significantly higher end-to-end delay performance compared to "Sc1" scenario where the video bearers have served with specific priority requirement.

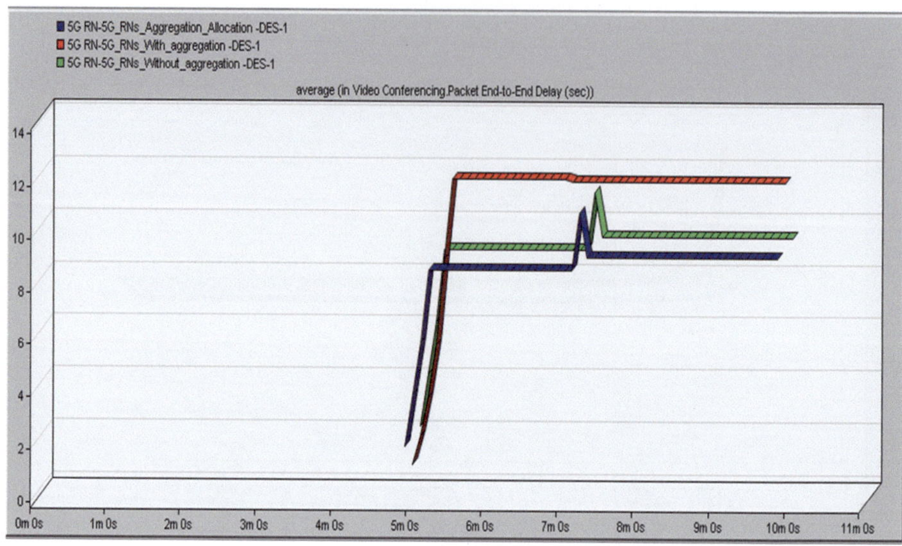

Fig. 1.16 Video average packets end-to-end delay (s)

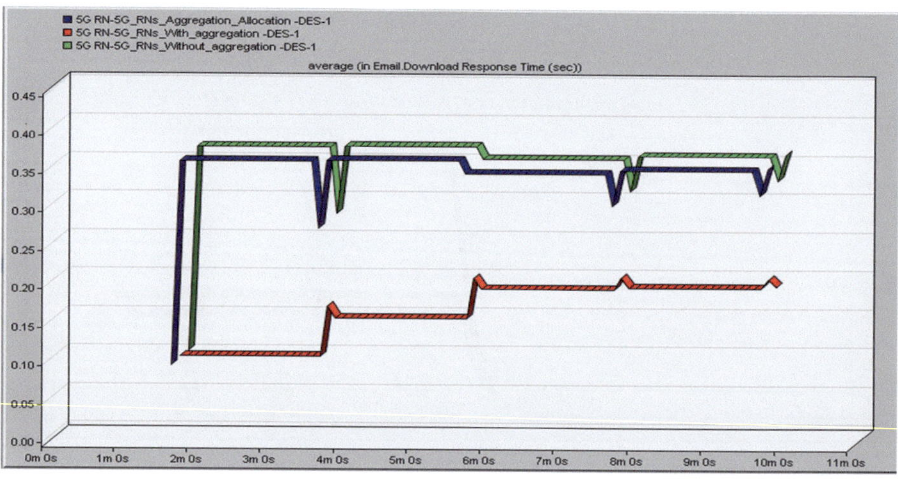

Fig. 1.17 Email average download response time (s)

Observing at the email bearers' results seen in Fig. 1.17., it can be observed that the email bearer has much better application performance when they not are allocated on a lower MAC QoS class as we can see in "Sc1" scenario. Mostly when it is not mixed with the FTP bearers and is allocated to a lower MAC QoS class than FTP. This is since of the QoS weight in "Sc1" scenario, which is considered the priority in a different level based on applications and smart systems need compared to "Sc2" and "Sc3" scenarios.

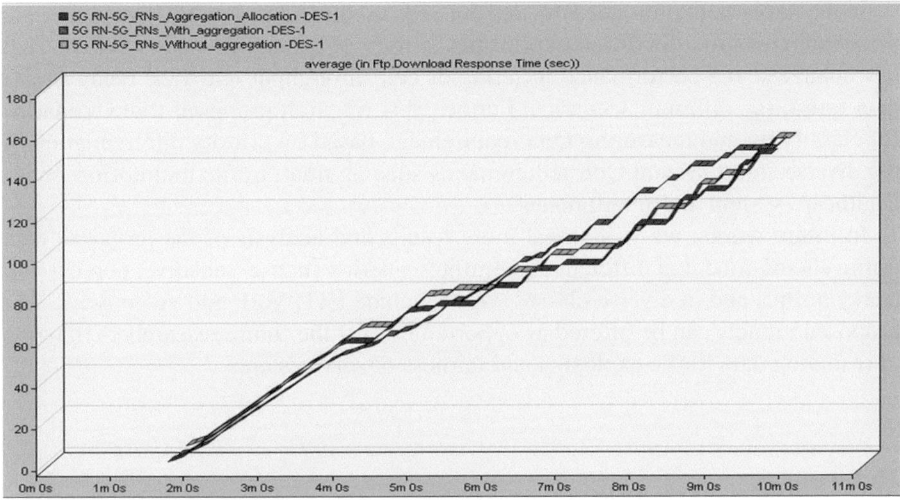

Fig. 1.18 FTP average download response time (s)

Lastly, the FTP bearer results are seen in Fig. 1.18. As already predictable, the FTP bearer performance is decreased when going from fully mixed scenario "Sc3" to fully separate one "Sc1," where the average file download time becomes improved. This is due to the FTP bearer that is allocated to the lowest MAC QoS class and is supported with low priority as compared to the other applications. However, offering the FTP bearer lower priority is realistic since FTP is not the real-time application and in real life, the FTP users are acceptable to wait a couple of more seconds for their files to be downloaded, while the same cannot be accepted when it comes to real-time applications such as video or VoIP.

1.6   Conclusion

This chapter proposed two models and algorithms. We proposed data traffic aggregation model and algorithm in fixed RNs for uplink in 5G cellular networks. It improves the radio resource utilization for smart systems over 5G mobile networks. It offers a maximum multiplexing gain in PDCP layer for data packets from the several smart devices along with considering diverse priorities to solve packets E2E delay. Also, in this chapter, we have presented a novel scheme for resource allocation in the 5G networks with network slicing. Our scheme is a heuristic-based prioritized resource allocation that takes into consideration both the inter- and the intra-slice priority and executes the resource allocation accordingly in order to meet the QoS requirements dictated by the service slice. Our scheme increases the QoE experienced by mobile UEs as well as allows a better management of network resources.

In the implementation, the RNs and 5G cells used to aggregate PRBs and allocate these radio resources in different priorities in form of slicing for smart devices. That has enhanced the performance in terms of cell throughput and E2E delay of 5G data traffic for different scenarios. Further, this research proposed three scenarios for classifying and measuring QoS requirement, based on priority differentiation of the diverse smart system QoS requirements such as smart traffic monitoring, smart healthcare system and smartphones.

In future works, we will reveal more results and analysis of the proposed data traffic slicing model in different data traffic scenarios such as sensitive, popular and heavy traffics and in diverse classes which include FTP, VoIP and video users. The proposed models can be offered as opportunities for the future researchers in terms of resolving data traffic explosion and fairness of services area.

References

1. Andrews, J. G., Buzzi, S., Choi, W., Hanly, S. V., Lozano, A., Soong, A. C. K., & Zhang, J. C. (2014). What Will 5G Be? *IEEE Journal on Selected Areas in Communications, 32*(6), 1065–1082.
2. Chen, M., Wan, J., & Li, F. (2012). Machine-to-machine communications: Architectures, standards, and applications. *KSII Transactions on Internet and Information Systems, 6*(2), 480–497.
3. Chen, K. C., & Lien, S. Y. (2014). Machine-to-machine communications: Technologies and challenges. *Ad Hoc Networks, 18*, 3–23.
4. Einsiedler, H. J., Gavras, A., Sellstedt, P., Aguiar, R., Trivisonno, R., & Lavaux, D. (2015). System design for 5G converged networks. In *2015 European Conference on Networks and Communications, EuCNC 2015* (pp. 391–396). Piscataway: IEEE.
5. Panwar, N., Sharma, S., & Singh, A. K. (2016). A survey on 5G: The next generation of mobile communication. *Physics Communication, 18*, 64–84.
6. Dighriri, M., Lee, G. M., Baker, T., & Moores, L. J. (2015). Measuring and classification of smart systems data traffic over 5G mobile networks. In B. Akhgar, M. Dastbaz, & H. Arabnia (Eds.), *Technology for smart futures*. Cham: Springer.
7. Zaki, Y., Zhao, L., Goerg, C., & Timm-Giel, A. LTE wireless virtualization and spectrum management. In *2010 3rd Joint IFIP Wireless and Mobile Networking Conference, WMNC 2010* (p. 2010). Piscataway: IEEE.
8. Liang, C., Yu, F. R., & Zhang, X. (2015). Information-centric network function virtualization over 5g mobile wireless networks. *IEEE Network, 29*(3), 68–74.
9. Zhu, K., & Hossain, E. (2016). Virtualization of 5G cellular networks as a hierarchical combinatorial auction. *IEEE Transactions on Mobile Computing, 15*(10), 2640–2654.
10. Costa-Perez, X., Swetina, J., Mahindra, R., & Rangarajan, S. (2013). Radio access network virtualization for future mobile carrier networks. *IEEE Communications Magazine, 51*(7), 27–35.
11. Rahman, M. M., Despins, C., & Affes, S. HetNet Cloud: Leveraging SDN & cloud computing for wireless access virtualization. In *2015 IEEE International Conference on Ubiquitous Wireless Broadband, ICUWB 2015* (p. 2015). Piscataway: IEEE.
12. Dighriri, M., Alfoudi, A. S. D., Lee, G. M., & Baker, T. (2017). Data traffic model in machine to machine communications over 5G network slicing. In *Proceedings – 2016 9th International Conference on Developments in eSystems Engineering, DeSE 2016* (pp. 239–244). Piscataway: IEEE.

13. Lee, Y. L., Chuah, T. C., Loo, J., & Vinel, A. (2014). Recent advances in radio resource management for heterogeneous LTE/LTE-A networks. *IEEE Communication Surveys and Tutorials, 16*(4), 2142–2180.
14. Abdalla, I., & Venkatesan, S. Remote subscription management of M2M terminals in 4G cellular wireless networks. In *Proceedings – Conference on Local Computer Networks, LCN, 2012* (pp. 877–885). Piscataway: IEEE.
15. Niyato, D., Hossain, E., Kim, D. I. K. D. I., & Han, Z. H. Z. (2009). Relay-centric radio resource management and network planning in IEEE 802.16 j mobile multihop relay networks. *IEEE Transactions Wireless Communications, 8*(12), 6115–6125.
16. Sui, Y., Vihriala, J., Papadogiannis, A., Sternad, M., Yang, W., & Svensson, T. (2013). Moving cells: A promising solution to boost performance for vehicular users. *IEEE Communications Magazine, 51*(6), 62–68.
17. Annunziato, A. (2015). 5G vision: NGMN – 5G initiative. *IEEE Vehicular Technology Conference, 2015*, (pp. 1–5). Boston; IEEE.
18. Iwamura, M. (2015). NGMN view on 5G architecture. *IEEE Vehicular Technology Conference, 2015*, (pp. 1–5). Boston; IEEE.
19. Alfoudi, A. S. D., Lee, G. M., & Dighriri, M. (2017). Seamless LTE-WiFi architecture for offloading the overloaded LTE with efficient UE authentication. In *Proceedings – 2016 9th International Conference on Developments in eSystems Engineering, DeSE 2016* (pp. 118–122). Piscataway: IEEE.
20. Kalil, M., Shami, A., & Ye, Y. (2014). Wireless resources virtualization in LTE systems. In *Proceedings – IEEE INFOCOM* (pp. 363–368). Piscataway: IEEE.
21. Muppala, S., Chen, G., & Zhou, X. (2014). Multi-tier service differentiation by coordinated learning-based resource provisioning and admission control. *Journal of Parallel and Distributed Computing, 74*(5), 2351–2364.
22. Hasan, M., Hossain, E., & Kim, D. I. (2014). Resource allocation under channel uncertainties for relay-aided device-to-device communication underlaying LTE-A cellular networks. *IEEE Transactions on Wireless Communications, 13*(4), 2322–2338.
23. Abu-Ali, N., Taha, A. E. M., Salah, M., & Hassanein, H. (2014). Uplink scheduling in LTE and LTE-advanced: Tutorial, survey and evaluation framework. *IEEE Communication Surveys and Tutorials, 16*(3), 1239–1265.
24. Zirong, G., & Huaxin, Z. (2009). Simulation and analysis of weighted fair queueing algorithms in OPNET. In *Proceedings – 2009 International Conference on Computer Modeling and Simulation, ICCMS 2009* (pp. 114–118). Piscataway: IEEE.
25. Zhu, K. & Hossain, E., 2016. Virtualization of 5G cellular networks as a hierarchical combinatorial auction. *IEEE Transactions on Mobile Computing, 15*(10), 2640–2654. Available at: http://ieeexplore.ieee.org/document/7348713/

Chapter 2
Challenges and Opportunities of Using Big Data for Assessing Flood Risks

Ahmed Afif Monrat, Raihan Ul Islam, Mohammad Shahadat Hossain, and Karl Andersson

2.1   Introduction

We never feel safe when calamity strikes, starting from the decimation of Hurricanes Katrina and Sandy to the Pacific earthquake, which moved from the main island of Japan, affected severely an atomic power plant in 2011 [1]. Hence, the devastation of both natural and human-caused disasters can appear to be both limitless and eccentric. Continuously, geological, hydrological, biological, and climatic variables are causing natural disasters resulting in catastrophic events that have disastrous consequences on environments and human social orders. Hazards can be categorized as geophysical and biological. Examples of geophysical hazards are earthquake and cyclone, while infestation and epidemic are the examples of biological hazards [2] (Fig. 2.1).

Among these various natural calamities, flood is considered one of the most catastrophic natural hazards, because its severity is very difficult to measure as different known and uncertain attributes are associated with it. Flood is not only responsible for the loss of human lives, but it also causes damage to the properties such as electric power transmission/generation lines, roads, transports, and crops. It is also responsible for severe water contamination and health hazards. Therefore, the assessment of the risk of flooding before its occurrence is crucial. This allows precautionary measures to be taken by warning the people living in the flood-

A. A. Monrat · R. U. Islam · K. Andersson (✉)
Department of Computer Science, Electrical and Space Engineering, Luleå University of Technology, Skellefteå, Sweden
e-mail: karl.andersson@ltu.se

M. S. Hossain
Department of Computer Science and Engineering, University of Chittagong, Chittagong, Bangladesh

© Springer International Publishing AG, part of Springer Nature 2018   31
M. M. Alani et al. (eds.), *Applications of Big Data Analytics*,
https://doi.org/10.1007/978-3-319-76472-6_2

Fig. 2.1 Flood-affected area of Chittagong, Bangladesh in 2017. (Source: http://bangladeshchronicle.net/2017/06/ctg-people-suffer-for-mora-fallout [3])

prone areas. Eventually, the people can be prepared themselves in advance to tackle the risk of flooding in an area. In this context, Big Data could play an important role in supporting the assessment of flood risks due to its capability to visualize, analyze, and predict the risks effectively. Its analytical methods are so robust that it can handle extremely large data chunks and is capable of processing complex mathematical computation to reveal patterns, trends, and associations in order to extract the values from the dataset to facilitate assessment or prediction more accurately [4]. Therefore, this book chapter presents the challenges and opportunities of Big Data in assessing the risk of flooding. This will be delineated by taking account of proposed system architecture as well as in the light of the existing research works on flood risk assessment.

2.2 Impact of Flood as a Natural Disaster

Flood is an example of catastrophic natural disaster since it brings huge amount of social, environmental, and economic impacts. Natural factors such as heavy rainfall, storm surge, tidal influences, and downstream water levels and human factors consisting of urbanization, deforestation, and obstruction of the channels in drainage systems are responsible for frequent flooding [5]. The vulnerability of people and infrastructure is decisive for the degree of harm and damage when floodwater physically encroaches on people and infrastructure [6]. Moreover, climate change, socioeconomic damage, increased population, and limited funds make this problem more critical. Flood has devastating impact on human lives as it causes loss of life and destruction of properties, crops, transport systems, and power grid, and

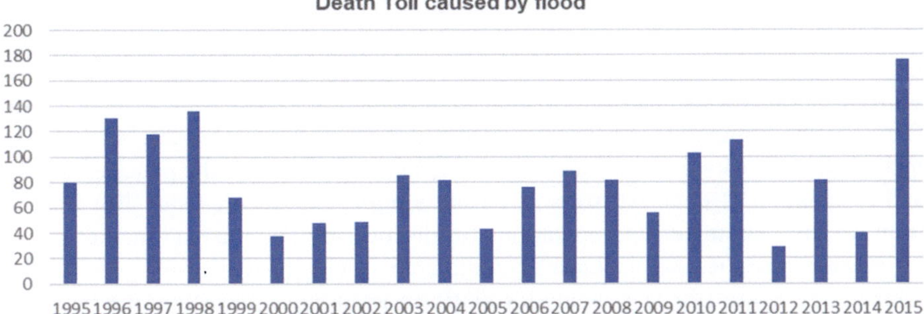

Fig. 2.2 Flood fatality statistics in the USA by the US Natural Hazard Statistics

the aftereffect can create health hazards and water contamination [7]. In the most recent decade of the twentieth century, flood is responsible for killing more than 100,000 people and influenced almost 1.4 billion individuals around the globe [9]. It is assessed that the yearly cost to the world economy because of flooding is around 50–60 billion US dollars [10]. As per an examination by the United Nations (UN), approximately 22,800 lives are affected significantly by flood in a year and caused an estimated damage of US\$ 136 billion to the Asian economy [10]. The damages suffered by the developing nations are five times higher per unit of GDP than those of developed nations [10]. More than 1200 people have died across India, Bangladesh, and Nepal and shut 1.8 million children out of school as a result of flooding in 2017 according to the reports of the guardian [11]. It is not possible to avoid flood risk entirely. However, different measures can be taken to prevent and mitigate the loss of flood; structural measures such as building dams or river dikes as well as nonstructural measures like flood forecasting, dissemination of flood warning, public participation, awareness, and institutional arrangement should be taken [8]. In order to reduce the damage due to flood, it is mandatory to monitor and evaluate the factors concerning floods to implement a system that can assess the risk of flood with highest accuracy. All sorts of information regarding floods such as weather forecasting, relevant data generating from different sensors, regional risk factors, and twitter feeds need to be taken into consideration for generating better risk assessment (Fig. 2.2).

2.3 Big Data for Flood Risk Management

Whenever an emergency requires a fast and effective reaction, it is frequently met with confusion and disorder. Communities from any part of the world are vulnerable to crisis, whether it's natural, human-actuated, or caused by different factors like flood [12]. However, flood risk management will always need some assistance in reaching out those who are affected and create a safer and secure environment for them, regardless of the gravity of this disaster. Therefore, Big Data has so much to promise in disaster management that is associated with flood. Big Data can deal with enormous volume of data which are coming from different sources in various formats [13]. Unlike traditional data processing approach, Big Data has the computing resources to process large and complex data in order to make better decision and provide valuable insight by assessing the patterns, trends, and association of data. All sorts of business organizations are using Big Data to improve their strategies and operations for discovering patterns and market trends to increase revenues, for instance, e-commerce service providers like Amazon or Alibaba are using Big Data analytics platform to monitor and study the behavior pattern of their consumers to find out new opportunities for customer satisfaction which ended up bringing more revenues for the company. Similarly, crisis response teams from different countries have turned their interest on Big Data in order to use its potential to come up with better prediction model for disaster like earthquakes, wildfires, storms, or floods [14]. The reason behind that is data are coming from different sources such as human, organizations, and machines while dealing with a natural disaster. By evaluating the data coming from social medias (Facebook or Twitter), sensors, satellite image, and disaster management organizations through API, many crises can be predicted before they occur which will give adequate time for evacuation of people and other crucial preparations. Moreover, Big Data is known for making unorganized sets of information into something comprehensive and meaningful. During a disaster, it becomes the most challenging task for the emergency response team to take appropriate measures within quick successions based on the inaccurate and incomplete information coming from various sources. For an example, crisis management team was struggling to reach people who needed help immediately during 2011 when Japan was struck by earthquake and tsunami [15]. Precisely under this kind of circumstances, Big Data can offer all sorts of aids for managing the disaster well.

2.3.1 How Can Big Data Help?

Big Data can increase social resilience to natural disaster by providing functionalities such as monitoring hazards, predicting exposure and vulnerabilities, managing disaster response, assessing the pliability of natural systems, as well as engaging communities throughout the disaster cycle [14]. Satellites, seismographs, and drone provide consistently enhancing remote-sensing abilities. Data that are coming from the smartphones and Twitter feeds create significant opportunities for monitoring hazards like floods or earthquakes [16]. Experts can identify geographical and infrastructural risks by using satellite images. Volunteers, as well as general people, can add ground level data by using crowdsourcing applications like OpenStreetMap or Ushahidi, for instance, people can inform their status during a flood or any kind of disaster to the authorities [17]. Social media can be monitored to study the behavior and movement of people after a natural calamity for guiding disaster response accordingly. To improve the agricultural interventions in developing nations, different sensors can be used in the field to reveal the quality of air and soil. By raising awareness among citizens, Big Data helps to build strong communities that can manage their natural system, strengthen infrastructure, and take effective decisions for a better future.

2.4 Opportunities of Big Data in Flood Risk Assessment

This segment will explain the scope of Big Data, which can bring new opportunities to improve the way flood risk managements are planned and executed. With Big Data, records of previous flood incidents such as fatality, the amount of damaged properties, rainfall during that period, infrastructures of the areas including coastal areas as well as cities, and drainage system can be analyzed properly [18]. It can also pick out the specific mobility support or resources that are needed by the inhabitants of a flood-affected area. Hence, identifying population hotspot gets easier with Big Data in order to provide real-time alarm and warnings to the residents when a disaster approaches (Fig. 2.3).

It helps to study future reactions of the people who are living in a specific zone and suffered tremendously by flood [19]. By using geographical image mapping technique, it is possible to map the risk zones in real time of a city or area, and viewers can observe the assessment through web services [20]. For instance, data scientist can extract detail information from local mobile network companies about how people reacted and responded to an emergency situation like flood. Moreover, Big Data makes a sort of spatial information framework in order to build the foundation which will make policies, protocols, and the trade of information as an ongoing priority. Such sharing of information makes new best-case situations to help both responders and survivors.

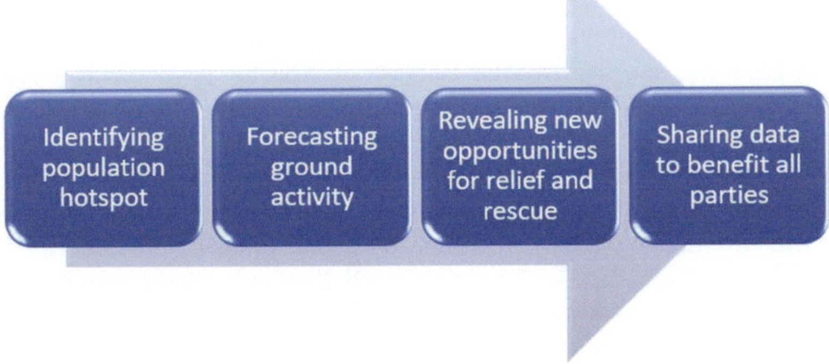

Fig. 2.3 Opportunities for Big Data for flood risk assessment

2.5 Challenges of Predicting Flood Risks

Despite of getting very promising results, Big Data needs to deal with some barriers, uncertainties, and risks associated with the assessment because of human and organizational capacity gaps along with the lack of access to internet and IT infrastructure especially in the developing countries. While implementing and scaling new approaches, Big Data is open to new risks due to specific technological, political, and economical obstacles [21]. For instance, the privacy and security of cell phone's data can be hampered by factors ranging from the large chunks of datasets due to uncertainty issues. Another example can be the analyses of social media data that works fine in developed countries; however, it may not be reliable in developing countries due to much thinner and more skewed base users. While leveraging Big Data to build resilience in complex and volatile environments, it is needed to be mindful about some factors such as constraints on data access and completeness; analytical challenges to actionability and replicability, for example, finding out the approaches to mitigate verification technique and sample bias correction methods; human and technology capacity gaps; and ethical and political risks [22]. Moreover, Big Data needs to comply with its major four Vs, that is, volume, velocity, variety, and veracity, as large amount of data needs to be processed that is coming from different sources in various formats with a high rate which can be unreliable and associated with lots of uncertainties [23]. During flood risk assessment, Big Data has to deal with various constraints in different phases like data acquisition, information extraction, data integration and analysis, data life cycle management, crowdsourcing, and disaster response recovery [24].

2.6 System Architecture Implementing Big Data

We can consider the system architecture in Fig. 2.4 to assess flood risk using Big Data. The system has three major parts: a data source, a Hadoop/Spark distributed server, and a web interface. The data source will provide data to the server which will perform computation regarding flood risks by using Big Data analytics platform, such as Apache Hadoop or Apache Spark, with a machine-learning approach called belief rule base (BRB), and finally through a web service, it will visualize the risk in a user interface.

It has three major data sources including human-generated data (Twitter, web traffic), sensors, and data that are coming from different organizations, for instance, GCM (Global Climate Model), real-time data, and imagery provided by NASA (Earthdata). All these data will be forwarded toward a server through an API gateway. The Big Data platform (Hadoop/Spark) will store this huge amount of data in different nodes by replicating a number of times in order to ensure the integrity of data providing fault tolerance, high availability, and scalability. YARN will be used as resource management layer that will distribute the computing tasks into different

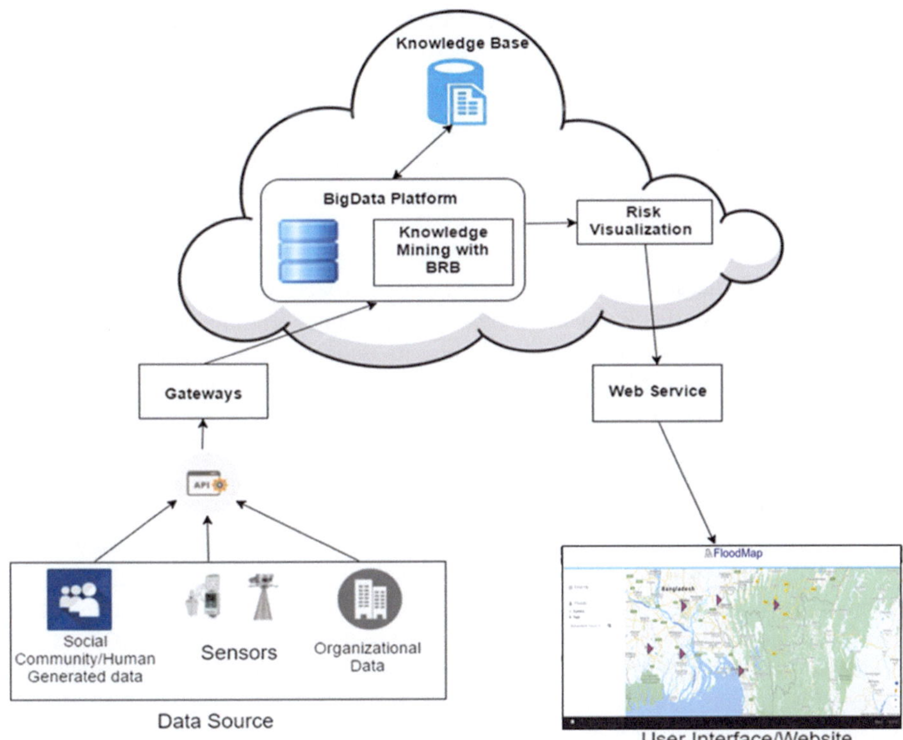

Fig. 2.4 System architecture

nodes. Then these data will be analyzed using BRB inference engine to get rid of the uncertainties associated with the data as well as to produce the risk assessment. The users can view the final assessment in a website for specific region. In addition, clients can also monitor the factors like water level, rainfall, and moisture of soil. The system will produce real-time assessment by computing stream of data coming from different sources. Therefore, crisis management teams, authorities, and people will be able to take proper precautions before flood occurs.

This system architecture will provide four major services: system management, real-time query, historical analytics, and forecast with warning. Through user management module, an administrator of the system can add or delete a user and manage user account permission. Real-time query provides information regarding the behavior, pattern, and trends of different sensors as well as the risks in real time. Historical analytics provides analysis of historical events, for instance, the variance and distribution of water in last couple of years, frequency of exceeding water level in a specific station per year, and infrastructural condition like drainage system in a region. Forecast and warning module are responsible for providing risk prediction.

2.6.1   Framework of the Assessment Model

A large number of predictive models are available for early prediction, real-time forecast of rainfall and water level in river stage, and flood risk assessment [25, 31]. Apparently, most of those models applied artificial neural network, support vector machine, dynamic Bayesian network, or a hybrid of these techniques. However, assessing flood risks or any kind of disaster before it occurs is quite complex and unreliable. Usually, expert systems are considered as suitable to handle problems of this nature rather than algorithmic approach. Expert systems have two major components: the knowledge base and inference engine. Knowledge base has the ability to solve a problem providing its underlying set of facts, rules, and assumptions, whereas inference engine helps the knowledge base to extract new information applying logical rules. Knowledge base can be constructed by using proportional logic (PL), first-order logic (FOL), or fuzzy logic (FL), yet these are not well equipped to capture uncertainty like ignorance, incompleteness, vagueness, and ambiguity with certain assurance [26].

Therefore, a recently developed belief rule-based inference methodology using the evidential reasoning (RIMER) approach is considered to develop the system architecture [27]. This methodology can address all sorts of uncertainties. The knowledge base is constructed with belief rule-based expert system (BRBES), while evidential reasoning (ER) works as an inference mechanism in this methodology [28]. Here, a rule base is constructed with belief degrees associated in all the possible consequences of a rule. The inference is implemented using the evidential reasoning approach that can handle different types and degrees of uncertainty associated with the flood risk assessment factors. Moreover, it is quite efficient to process large stream of sampled data that are coming from Hadoop distributed file system (Fig. 2.5).

The inference procedure in BRB consists of four phases: input transformation, rule activation and weight calculation, belief degree update, and rule aggregation using evidential reasoning. In the first phase, input data will be transformed into a distribution of referential values of an antecedent attribute [29]. For example, if the precipitation rate of rainfall is 3 mm/h and it comes under the threshold between heavy rain 16 mm/h and moderate rain 4 mm/h, it will be transformed into three referential values (light rain [10%], moderate rain [78%], and heavy rain [12%]) according to the preset thresholds. The next phase will provide an activation weight and matching degree to the input. The belief degree associated with each rule in the rule base should be updated when an input data for any of the antecedent is ignored or missing [30]. For instance, if the input value of the factor "river water level" is somehow ignored, then the belief degree will be updated. Finally, rule aggregation method associates all the rules and belief degrees and performs computation to find out if the final assessment will be converted again as a crisp value using different referential scores [31].

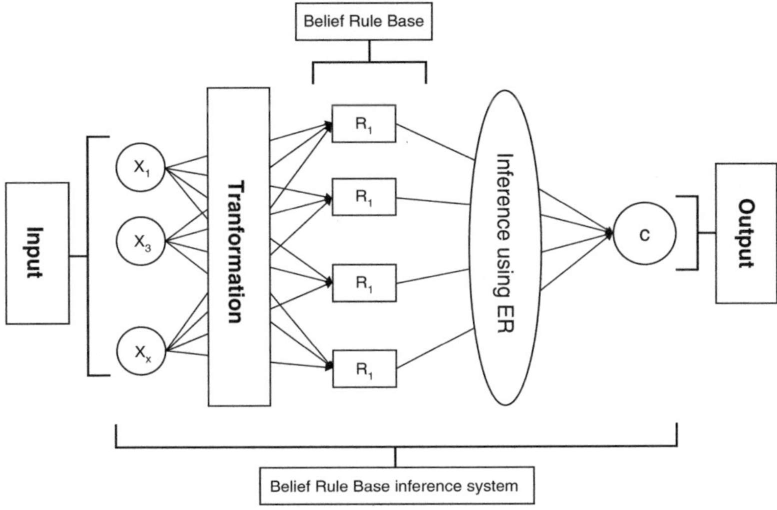

Fig. 2.5 BRBES inference mechanism

2.7   Current Research on Flood Prediction Using Big Data

Use Case 1 A group of scientists (De Groeve, Kugler, and Brakenridge) built a real-time map of location, timing, and impact of floods by combining information related to flood from twitter and satellite observations [32]. It is possible to update the map constantly and can be accessed online.

Use Case 2 In the Netherlands, the government has started experimenting with how machine learning may help strengthen preparedness to future floods, where the vast majority of the population lives in flood-prone areas (Fig. 2.6) [33].

Use Case 3 An early warning system has been developed by the New South Wales state emergency service in Australia. It takes meteorological dataset such as data from flood plain and historical data information from various databases to perform predictive analysis of floods in different region [34].

Use Case 4 A social media analytics platform named FloodTags was deployed for extracting information from twitter (Fig. 2.6). It has the functionality to perform filtering, visualization, and mapping social media content based on location and keywords [35]. Besides, it also provides a service through microwave satellite observations for identifying inundated areas rapidly. The approach has been used in the Philippines and Pakistan as case studies which later proved to be a great success monitoring large floods in densely populated areas.

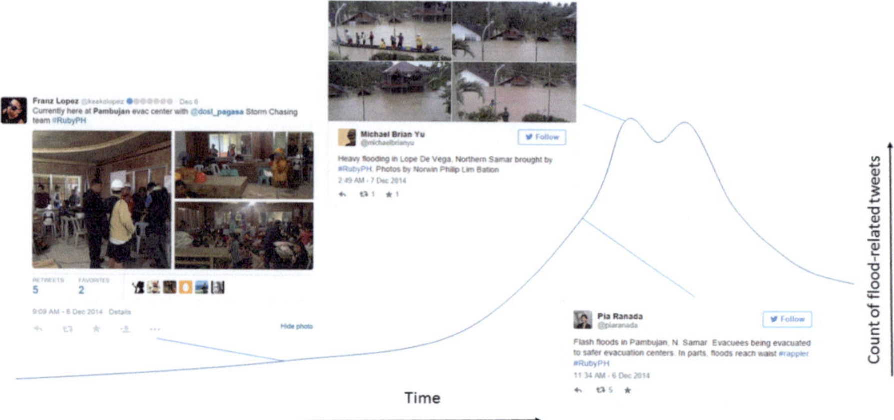

Fig. 2.6 Monitoring flood risks using Big Data Analytics platform

2.8 Conclusion

As the access to datasets is expanding rapidly due to smartphones and various sensors, the utility and potential of Big Data for disaster management are growing [36]. However, depending on how Big Data is used in different disaster phases, the hype can be turned into an improved disaster risk management. Due to serious policy, operational, and even philosophical issues, the integration of Big Data into existing workflows has become very challenging. As crisis response teams are strengthening their ties with the formal emergency management sector, it is expected that Big Data technology can offer aids which will suit the needs of responders. Big Data is more than just data, as it comprises new techniques to extract information from a large volume of data which can possibly change the future of our next generation. As we need all sorts of relevant data in order to assess flood risks with higher precision, it is mandatory to build a resilience system with open and public data and data centers for cloud computing. In addition, building partnerships with private companies and municipalities, in order to increase data access and provide research grants to rigorously test new algorithms with university faculties and students, is crucial in advancing Big Data applications. To mitigate the losses caused by flood, Big Data analytic teams in disaster risk management should also involve local participation as crowdsourcing. Finally, it is safe to say that Big Data is all about making proper decisions by analyzing data which will influence precautions, response, and quick recovery, the necessary ingredients for making disaster management effective.

Acknowledgment This research has been supported by Pervasive Computing and Communications for Sustainable Development (PERCCOM) and the Swedish Research Council under grant 2014-4251. PERCCOM is a joint master degree program funded by a grant from the European Union's Erasmus Mundus program. The authors would like to acknowledge the European Union. We also thank all the PERCCOM faculties and students from around the world.

References

1. Tominaga, T., Hachiya, M., Tatsuzaki, H., & Akashi, M. (2014). The accident at the fukushima daiichi nuclear power plant in 2011. *Health physics, 106*(6), 630–637.
2. Gill, J. C., & Malamud, B. D. (2014). Reviewing and visualizing the interactions of natural hazards. *Reviews of Geophysics, 52*(4), 680–722.
3. CTG people suffer for Mora fallout. Retrieved on Tuesday, November 14, 2017, http://bangladeshchronicle.net/2017/06/ctg-people-suffer-for-mora-fallout
4. De Mauro, A., Greco, M., & Grimaldi, M. (2016). A formal definition of Big Data based on its essential features. *Library Review, 65*(3), 122–135.
5. Morita, M. (2014). Flood risk impact factor for comparatively evaluating the main causes that contribute to flood risk in urban drainage areas. *Water, 6*(2), 253–270.
6. Kezia, S. P., & Mary, A. V. A. (2016). Prediction of rapid floods from big data using map reduce technique. *Global Journal of Pure and Applied Mathematics, 12*(1), 369–373.
7. Doocy, S., Daniels, A., Packer, C., Dick, A., & Kirsch, T. D. (2013). The human impact of earthquakes: A historical review of events 1980–2009 and systematic literature review. *PLoS currents, 5*.
8. Jonkman, S. N. (2005). Global perspectives on loss of human life caused by floods. *Natural hazards, 34*(2), 151–175.
9. Shrestha, A. B., Shah, S. H., & Karim, R. (2008). *Resource manual on flash flood risk management*. Kathmandu: Internat. Centre for Integrated Mountain Development, ICIMOD.
10. Shrestha, M. S., & Takara, K. (2008). Impacts of floods in south Asia. *Journal of South Asia Disaster Study, 1*(1), 85–106.
11. South Asia floods: Mumbai building collapses as monsoon rains wreak havoc. Retrieved on August 31, 2017, https://www.theguardian.com/world/2017/aug/31/south-asia-floods-fears-death-toll-rise-india-pakistan-mumbai-building-collapses
12. Seck, P. (2007). Links between natural disasters, humanitarian assistance and disaster risk reduction: A critical perspective. *Occasional Paper, 15*.
13. Gandomi, A., & Haider, M. (2015). Beyond the hype: Big data concepts, methods, and analytics. *International Journal of Information Management, 35*(2), 137–144.
14. Emmanouil, D., & Nikolaos, D. (2015). *Big data analytics in prevention, preparedness, response and recovery in crisis and disaster management*. In The 18th International Conference on Circuits, Systems, Communications and Computers (CSCC 2015), Recent Advances in Computer Engineering Series, Vol. 32, pp. 476–482.
15. Zaré, M., & Afrouz, S. G. (2012). Crisis management of Tohoku; Japan earthquake and tsunami, 11 March 2011. *Iranian Journal of Public Health, 41*(6), 12.
16. Yusoff, A., Din, N. M., Yussof, S., & Khan, S. U. (2015, December). *Big data analytics for Flood Information Management in Kelantan, Malaysia*. In Research and Development (SCOReD), 2015 IEEE Student Conference on (pp. 311–316). IEEE.
17. Xu, Z., Liu, Y., Yen, N., Mei, L., Luo, X., Wei, X., & Hu, C. (2016). Crowdsourcing based description of urban emergency events using social media big data. *IEEE Transactions on Cloud Computing*.
18. Labrinidis, A., & Jagadish, H. V. (2012). Challenges and opportunities with big data. *Proceedings of the VLDB Endowment, 5*(12), 2032–2033.
19. Chen, C. P., & Zhang, C. Y. (2014). Data-intensive applications, challenges, techniques and technologies: A survey on Big Data. *Information Sciences, 275*, 314–347.
20. Wang, S., & Yuan, H. (2014). Spatial data mining: A perspective of big data. *International Journal of Data Warehousing and Mining (IJDWM), 10*(4), 50–70.
21. Bello-Orgaz, G., Jung, J. J., & Camacho, D. (2016). Social big data: Recent achievements and new challenges. *Information Fusion, 28*, 45–59.
22. Jin, X., Wah, B. W., Cheng, X., & Wang, Y. (2015). Significance and challenges of big data research. *Big Data Research, 2*(2), 59–64.

23. Chen, H., Chiang, R. H., & Storey, V. C. (2012). Business intelligence and analytics: From big data to big impact. *MIS Quarterly, 36*(4).
24. Demchenko, Y., Zhao, Z., Grosso, P., Wibisono, A., & De Laat, C., 2012, December. Addressing big data challenges for scientific data infrastructure. In Cloud Computing Technology and Science (CloudCom), 2012 IEEE 4th International Conference on (pp. 614–617). IEEE.
25. Ul Islam, R., Andersson, K., & Hossain, M.S., 2015, December. A web based belief rule based expert system to predict flood. In *Proceedings of the 17th International conference on information integration and web-based applications & services* (p. 3). ACM.
26. Alharbi, S. T., Hossain, M. S., & Monrat, A. A. (2015). A Belief Rule Based Expert System to Assess Autism under Uncertainty. In Proceedings of the World Congress on Engineering and Computer. *Science, 1.*
27. Yang, J. B., Liu, J., Wang, J., Sii, H. S., & Wang, H. W. (2006). Belief rule-base inference methodology using the evidential reasoning approach-RIMER. *IEEE Transactions on systems, Man, and Cybernetics-part A: Systems and Humans, 36*(2), 266–285.
28. Yang, J. B. (2001). Rule and utility based evidential reasoning approach for multiattribute decision analysis under uncertainties. *European journal of operational research, 131*(1), 31–61.
29. Hossain, M. S., Zander, P. O., Kamal, M. S., & Chowdhury, L. (2015). Belief-rule-based expert systems for evaluation of e-government: A case study. *Expert Systems, 32*(5), 563–577.
30. Wang, Y. M., Yang, J. B., & Xu, D. L. (2006). Environmental impact assessment using the evidential reasoning approach. *European Journal of Operational Research, 174*(3), 1885–1913.
31. Hossain, M. S., Monrat, A. A., Hasan, M., Karim, R., Bhuiyan, T. A., & Khalid, M. S. (2016, May). *A belief rule-based expert system to assess mental disorder under uncertainty.* In Informatics, Electronics and Vision (ICIEV), 2016 5th International Conference on (pp. 1089–1094). IEEE.
32. Kugler, Z., & De Groeve, T. (2007). *The global flood detection system.* Luxembourg: Office for Official Publications of the European Communities.
33. Pyayt, A. L., Mokhov, I. I., Lang, B., Krzhizhanovskaya, V. V., & Meijer, R. J. (2011). Machine learning methods for environmental monitoring and flood protection. *World Academy of Science, Engineering and Technology, 78*, 118–123.
34. Zerger, A. (2002). Examining GIS decision utility for natural hazard risk modelling. *Environmental Modelling & Software, 17*(3), 287–294.
35. Jongman, B., Wagemaker, J., Romero, B. R., & de Perez, E. C. (2015). Early flood detection for rapid humanitarian response: Harnessing near real-time satellite and Twitter signals. *ISPRS International Journal of Geo-Information, 4*(4), 2246–2266.
36. Tellman, B., Schwarz, B., Burns, R., & Adams, C. UN Development Report 2015 Chapter Disaster Risk Reduction Big Data in the Disaster Cycle: Overview of use of big data and satellite imaging in monitoring risk and impact of disasters.

Chapter 3
A Neural Networks Design Methodology for Detecting Loss of Coolant Accidents in Nuclear Power Plants

David Tian, Jiamei Deng, Gopika Vinod, T. V. Santhosh, and Hissam Tawfik

3.1 Introduction

Nuclear power plants (NPP) life management is concerned with monitoring the safety and the conditions of the components of a NPP and the maintenance of the NPP in order to extend its lifetime. It is crucial to regularly monitor the safety of the components of a NPP to detect as early as possible any serious anomalies which would potentially cause accidents. When an accident is predicted to occur or occurring, the plant operator must take necessary actions as quickly as possible to safeguard the NPP, which involves complex judgements, making trade-offs between demands and requires a lot of expertise to make critical decisions. It is commonly believed that timely and correct decisions in these situations could either prevent an event from developing into a severe accident or mitigate the undesired consequences of an accident. As nuclear power plants become more advanced, their safety monitoring approaches grow considerably. Current approaches include nuclear reactor simulators, safety margin analysis [1, 2], probabilistic safety assessment (PSA) [1, 3] and artificial intelligence (AI) methods such as neural networks [4–7]. Nuclear reactor simulators such as RELAP5-3D [8] simulate the dynamics of a NPP in accidental scenarios and generate transient datasets of reactors. Safety margin analysis analyses the values of the safety parameters of a reactor and triggers an alert to the plant operator if the safety margin falls below a minimum safety margin.

D. Tian · J. Deng (✉) · H. Tawfik
School of Computing, Creative Technologies and Engineering, Leeds Beckett University, Leeds, UK
e-mail: J.Deng@leedsbeckett.ac.uk

G. Vinod · T. V. Santhosh
Reactor Safety Division, Bhabha Atomic Research Centre, Mumbai, India

© Springer International Publishing AG, part of Springer Nature 2018
M. M. Alani et al. (eds.), *Applications of Big Data Analytics*,
https://doi.org/10.1007/978-3-319-76472-6_3

PSA computes the probability of occurrence of accidents based on the probabilities of the component failures which cause the accidents. These approaches are often used to together to safeguard NPPs.

Machine learning is a subfield of AI and is the study of the algorithms such as neural networks that learn from data to make decisions. In machine learning, the amount of training patterns can critically affect the predictive performances of neural networks. When the training set contains limited amount of training patterns, the prediction performance of models can be improved by adding new patterns to the training set. Linear interpolation method [9] is a well-known technique for generating artificial data points. Whereby, new data points are introduced from a set of known data points by creating straight lines to connect the known data points and taking the points (new data points) on the lines. Neural networks can be trained on transient datasets of a NPP to detect LOCA of the NPP. However, the transient datasets exhibit big data characteristics and designing an optimised neural network by exhaustive training all possible neural network architectures on big data can be very time-consuming because there exist a large number of possible neural network architectures for big data. The objective of this work is to propose a fast methodology to design neural networks using a transient dataset of IHs and a new break size dataset generated using linear interpolation to identify break sizes of inlet headers (IHs) of a pressurised heavy water reactor (PHWR) [5, 10]. We consider this work as an engineering application of predictive data analytics for which neural networks are used as the primary tool.

This chapter is organised as follows: Section 3.2 reviews the current approaches for monitoring the safety of nuclear power plants; Sect. 3.3 describes the large break LOCA and the generation of a transient dataset of IHs using RELAP5-3D; Sect. 3.4 proposes the methodology of training neural networks for LOCA detection; Sect. 3.5 presents the results of applying the proposed methodology to LOCA detection; Sect. 3.6 discusses the results; and conclusions and future work are presented in Sect. 3.7.

3.2 Approaches for Monitoring the Safety of Nuclear Power Plants

Nuclear reactor simulators such as Reactor Excursion and Leak Analysis Program (RELAP5-3D) [8] and Modular Accident Analysis Program (MAAP5) [11] have been used to support human operators analysing the safety of NPPs. RELAP5-3D, developed by the Idaho National Laboratory of the USA, is a dedicated tool to analyse transients and accidents in pressurised water reactors (PWR). One key feature of RELAP5-3D is its multidimensional thermal-hydraulic capability which simulates the dynamics of a water-cooled reactor in great details to enable detailed analysis of severe accidents such as LOCA. Therefore, RELAP5-3D has been widely used for simulating accidental scenarios of PWRs. The safety indicators of a reactor are the various parameters such as the steam flow rates of its components, the pressure and the temperature of its components. The simulators use thermal-hydraulic codes to simulate the dynamics of the parameters of reactors during accidental scenarios such as LOCA. MAAP5 owned by Electric Power Research Institute (EPRI) simulates the dynamics of various types of water-cooled reactors including advanced light water reactors (ALWRs), boiling water reactors (BWRs) and PWRs during severe accidents [11].

Safety margin analysis [1, 2] has been used to safeguard the NPP. Safety margin [2] is the difference between the values reached by the safety parameters of a reactor during accidental scenarios and the preset thresholds that must not be exceeded in order to maintain the safety of the NPP. The larger the safety margin, the safer the NPP is and vice versa. Regulatory bodies have specified minimum safety margins [2] beyond which a NPP is not safe to operate. To compute the safety margins of a reactor, wireless sensors are attached to the components of a reactor to collect the real-time data of the safety parameters [10]. The values of the safety parameters and their thresholds can also be retrieved from a simulation of an accidental scenario to compute safety margins. If a safety margin falls below the respecified minimum safety margin, an alert is generated and passed to the plant operator.

Probabilistic safety assessment (PSA) methods [1, 3] such as event trees (ET) [1, 3, 12], dynamic event trees (DET) [13, 14] and fault trees (FT) [1, 3, 12] have been developed to compute the probability of occurrence of accidental scenarios so that the operator can locate the component failures related to the accidental scenarios and take the necessary actions. An accident is the outcome of a sequence of component failures which are triggered by an initiating event such as station black out (SBO) [1]. The initiating event is at the root (top node) of an ET. Each branch of the tree represents the probability of occurrence or non-occurrence of a component failure. A bottom branch is the probability of occurrence or non-occurrence of an outcome. DET is similar to event trees except that DET uses a time-dependent model to determine the timing and the sequence of the responses of the system under analysis [13, 14]. The time-dependent model of the system accounts for different timing, order and magnitude of the possible failure events. Therefore, a DET analysis of a system covers a much larger set of possible scenarios than that of an ET analysis. Fault trees are used to identify the occurrence of the initiating failure which has caused an accident [12]. With the accident at the root, walking backwards from the root, all the possible failures causing the accident are identified, and their probabilities of occurrences are represented as branches of the root. The tree grows by adding new branches under each lowest-level branch. The bottom branch is the probability of occurrence of an initiating failure.

Artificial intelligence approaches such as robotics [15, 16], neural networks [4–7] and fuzzy systems [17, 18] have gained considerable attention in detecting failure and accidents of nuclear systems over the past decade. Continual inspection of critical components such as the primary heat transport (PHT) is crucial to maintaining the safety of the NPP. However, human inspection is dangerous and difficult due to the hazardous environment and geometric restraints of the components [15, 16]. Inspection robotics has been used as an alternative to human inspectors to determine early warning of component failure and prevent possible nuclear accidents. A snake-armed inspection robot attached with high resolution camera has been used to examine the conditions of the PHT pipes of a reactor under hazardous environments which are too dangerous for human inspectors [15]. Successful predictive models have been developed by training and testing neural networks and fuzzy systems on transient datasets generated using RELAP5-3D and MAAP5 simulators. Na MG et al. [4] generated transient data of IHs using MAAP4 code and trained neural networks on the transient data to detect LOCA in an advanced power reactor 1400 (APR1400). Zio E et al. [17] applied fuzzy similarity analysis approaches to detecting the failure modes of nuclear systems. Wang WQ et al. [19] developed a neuro-fuzzy system to predict the fuel rod gas pressure based on cladding inner surface temperatures in a loss-of-coolant accident simulation. Baraldi P et al. [18] proposed an ensemble of fuzzy C-mean classifiers to identify the faults in the feed water system of a boiling water reactor. Wei X et al. [20] developed self-organising radial basis function (RBF) networks to predict fuel rod failure of nuclear reactors. Souza [21] developed a RBF network capable of online identifying the accidental dropping of the control rod at the reactor core of a pressurised water reactor. Secchi P et al. [22] developed bootstrapped neural networks to estimate the safety margin on the maximum fuel cladding temperature reached during a header blockage accidental scenario of a nuclear reactor. Back J et al. [23] developed a cascaded fuzzy neural network using simulation data of the optimised power reactor 1000 to predict the power peaking factor in the reactor core in order to prevent nuclear fuel melting accidents. Guimarães A and Lapa C [24] used an adaptive neural fuzzy inference system (ANFIS) to detect the cladding failure in fuel rods of a nuclear power plant. Santhosh et al. [5] trained a neural network on a transient dataset generated using RELAP5-3D to detect the size of a break, the location of the break in the PHT with the availability of the emergency core cooling system (ECCS) which automatically shuts down the reactor to prevent a subsequent accident.

3.3   Large Break Loss of Coolant Accidents of a PHWR

This work uses RELAP5-3D to simulate the dynamics of the parameters of a PHWR in LOCA scenarios and generate transient datasets for training neural networks to detect the break sizes of the IHs during LOCA scenarios. The PHWR [5, 10] is a main nuclear reactor currently in operation in Indian. A LOCA is caused by a large break of the IHs of the primary heat transport system (PHT) (Fig. 3.1) of a PHWR as follows. When large breaks of inlet headers of the PHT occur, the system depressurises rapidly which causes coolant voiding into the reactor core. This coolant voiding into the core causes positive reactivity addition and consequent

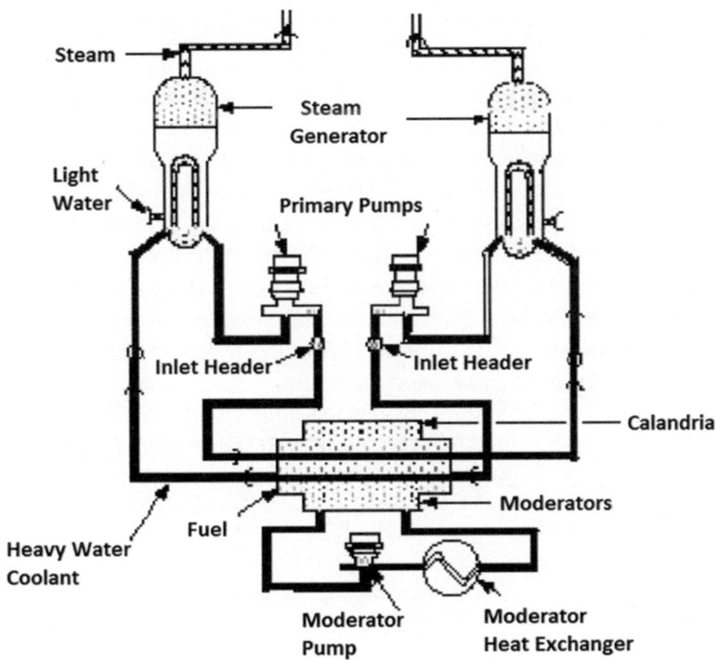

Fig. 3.1 The PHT of a PHWR [4]

power rise. Then, the emergency core cooling system automatically shuts down the reactor to keep the NPP safe. During an occurrence of a break, transient data such as the temperature and pressure of the IHs can be collected during a short time period to detect the sizes of the break using neural networks. The break size is defined as the percentage of the cross-sectional area of an IH. The break size is between 0% (no break) and 200%, i.e. double cross-sectional areas of an IH (a complete rupture of the IH). It is infeasible to generate all possible break sizes. In this study, a transient dataset consisting of the 6 break sizes 0%, 20%, 60%, 100%, 120% and 200% was generated using RELAP5-3D. The break sizes of 20% or greater are considered as large breaks. For each break size, the 37 signals used by Santhosh et al. [5] were collected at various parts of the PHT over 60 seconds using RELAP5-3D under the assumption that this time duration is sufficient to identify large break LOCA in IH. The 37 signals are measurements of the flow rate, the temperatures and the pressures of the various parts of the PHT. For each break size, the signals were measured at 541 time instants within a 60s duration. Each break size class of the transient dataset consists of 541 instances (observations) and 37 features (signals). The transient dataset is a 3246 × 38 matrix with the last column representing the break size (the output).

3.4 The Neural Networks Training Methodology

The proposed methodology consists of three stages (Fig. 3.2). In the first stage, a number of 1-hidden layer MLP architectures are created empirically. Then, each architecture is trained and tested a number of times using the transient dataset to select an optimised 1-hidden layer MLP (Sect. 3.4.3). In the second stage, a number of 2-hidden layer MLP architectures with equal number of nodes in each hidden layer are created based on the number of the weights of the optimised 1-hidden layer MLP. Then, each 2-hidden layer architecture is trained and tested a number of times using the transient dataset to select an optimised 2-hidden layer MLP (Sect. 3.4.4). In the third stage, the break sizes not present in the transient dataset are generated using linear interpolation method; then, the optimised 2-hidden layer MLP is trained and tested iteratively on the transient dataset added with the linear interpolation dataset to select an optimised MLP. The initial weights of the network in a training-testing iteration are set to the weights of the trained network in the last training-testing iteration (Sect. 3.4.5).

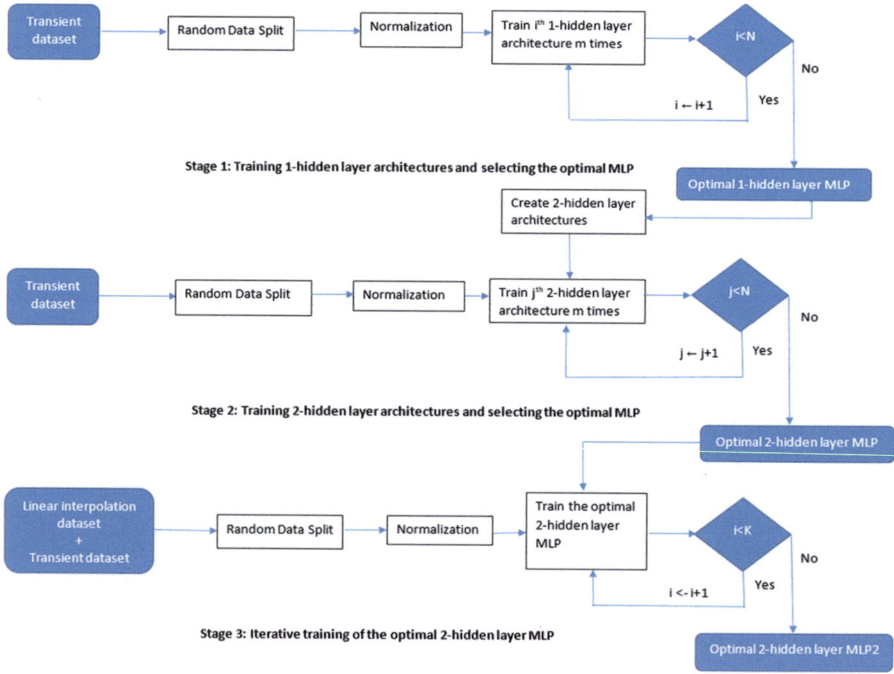

Fig. 3.2 The proposed neural networks training methodology

3.4.1   Performance Measures

The following performance measures are used to evaluate the performances of the neural networks in detecting LOCA:

- Root mean square error (RMSE) of a break size:

$$\mathrm{RMSE}_K = \sqrt{\mathrm{MSE}_K}$$

and

$$\mathrm{MSE}_K = \frac{\sum_{i=1}^{M} (O_i - T_i)^2}{M}$$

where k is a break size, i.e. $k = 0\%, 20\%, 60\%, 100\%, 120\%$ or 200%; M is the number of the patterns of break size k in the test set; i is the ith pattern of break size k in the test set; O_i is the output of the network for the ith pattern; and T_i is the break size target of the ith pattern.

- Mean RMSE $= \frac{\sum_K \mathrm{RMSE}_K}{N}$
  where N is the number of the different break sizes in the test set; in this study, $N = 6$.
- Standard deviation of RMSEs $= \sqrt{\frac{\sum_K \left(\mathrm{RMSE}_K - \overline{\mathrm{RMSE}}\right)^2}{N-1}}$
  where $\overline{\mathrm{RMSE}}$ is the mean RMSE.

The RMSE of a break size measures the performance of a network in detecting that specific break size. The mean RMES measures the average performance of a network in detecting break sizes. The standard deviation of RMSEs measures the stability/variation of the performance of a network in detecting break sizes.

3.4.2   Random Data Split and Normalisation of the Transient Dataset

The transient data is randomly split into a 50% training set, a 25% validation set and a 25% test set using the random subsampling with no replacement method [25–28]. This creates a balanced training set, a balanced validation set and a balanced test set which are non-overlapping subsets of the transient dataset. Each break size class of the training set consists of 270 instances which are uniformly drawn at random from the transient dataset. The advantage of using a balanced training set is that this would ensure that the trained neural network would make unbiased estimation of the different break sizes in the test set [25, 27]. In contrast, a network trained on an imbalanced training set would tend to output the break size target corresponding to the majority class of the training set which leads to poor performance of the network. The 37 inputs and the break size targets of the training set are rescaled to the interval $[-1, 1]$ using min-max normalisation before training neural networks. When testing the trained networks, the outputs of the networks for the test set are transformed back to the target break size range $[0\%, 200\%]$ by inversing the min-max normalisation calculation.

3.4.3 Training of 1-Hidden Layer MLPs and Selection of the Optimised 1-Hidden Layer MLP

Firstly, a number of single hidden layer multilayer perceptrons (MLPs) were trained on the training set and evaluated on the test set. The validation set was used to validate the performance of the network during training. Each MLP has 37 inputs, a number of hidden nodes and 1 output node. Each hidden node is a logistic sigmoid function [7]:

$$\text{Logistic}\,(a) = \frac{1}{\left[1 + e^{-a}\right]} \tag{3.1}$$

and

$$a = \sum_i x_i w_i + b \tag{3.2}$$

where x_i is the ith input, w_i is ith weight and b is the bias of the hidden layer. The output node O of each MLP is the linear function [7]:

$$O = \sum_i x_i w_i + b_o \tag{3.3}$$

where b_o is the bias of the output node. Six 1-hidden layer MLPs with 10, 12, 15, 18, 20 and 22 hidden nodes were trained, respectively, using the Levenberg-Marquardt algorithm [29, 30] with setting of the maximum epoch to 1000 and that of the learning rate to 0.001. The criterion to terminate training is that training stops if the validation error does not improve for six consecutive epochs. Using this termination criterion would find an optimised set of weights and would prevent the network from overfitting the training set which could be caused by training for longer time. Each of the six 1-hidden layer MLPs was trained five times on the training set giving a total of 30 MLPs (Fig. 3.2). The optimised network among the 30 networks is the one with the smallest mean RMSE and the smallest number of nodes because the network with the smallest size and the highest predictive performance has the best generalisation performance on unseen data.

3.4.4   Training of 2-Hidden Layer MLPs and Selection of the Optimised 2-Hidden Layer MLP

A 2-hidden layer MLP [29, 30] is a universal approximator which can approximate any non-linear continuous function to any degree of accuracy. Networks with excessive number of nodes and weights may overfit the training set and have poor generalisation performance. Our optimised 1-hidden layer MLP has 18 hidden nodes and 684 weights. The number of the weights of the optimised 1-hidden layer MLP was used as a guidance to determine the number of the weights of a 2-hidden layer MLP based on the assumption that 2-hidden layer MLPs with similar number of weights as the optimised 1-hidden layer MLP have high generalisation performance. The lower and the upper bounds (LB and UB) of the number of the weights of a 2-hidden layer MLP were obtained using the heuristic rules LB = 684-L and UB = 684 + M where L and M are values set by the user. Setting L to 200 and M to 100 gives a LB of 484 and an UB of 784. The formula relating the number of weights W and the number of nodes H of each hidden layer of a 2-hidden layer MLP with equal number of nodes in each hidden layer is $W = 37 \times H + H^2 + H$ which can be rearranged to the following quadratic equation:

$$H^2 + 38H - W = 0. \tag{3.4}$$

Setting W to 484 and 784, respectively, the positive solutions to Eq. (3.4) are H = 10 and H = 15. Therefore, 6 2-hidden layer MLPs with 10, 11, 12, 13, 14 and 15 nodes in each hidden layer were trained, respectively. Each of the 6 2-hidden layer MLPs was trained 5 times on the training set giving a total of 30 networks. The optimised network among the 30 trained networks is the network with the smallest mean RMSE and the smallest number of nodes.

3.4.5   Training the Optimised 2-Hidden Layer MLP on Linear Interpolation Dataset and Transient Dataset

Linear interpolation [9] is a method of constructing new data points within the range of a set of known data points by fitting straight lines using linear polynomials. Having obtained the optimised 2-layer MLP, the break sizes 2.5%, 5%, 7.5%, 10%, 12.5%, . . ., 195% and 197.5% which are missing in the transient dataset, were generated using linear interpolation. For each missing break size, 541 instances were generated giving a total of 40,575 instances. The transient dataset and the break size dataset generated by linear interpolation were merged into a dataset containing 43,821 instances. Thereafter, the optimised 2-layer MLP was trained and tested iteratively 100 times on the merged dataset to obtain a MLP with better performance than the optimised 2-layer MLP. During each training-testing process, the merged data was randomly split into a 50% training set, a 25% validation set and a 25% test set; then, the weights of the network trained in the previous training-testing process were used as the initial weights of the current training-testing process before training began. This would give faster training speed than setting the initial weights to random values because each training process started at a minimum point on the error surface and stopped at another minimum point in the local region of the minimum point of the last training process. The mean RMSE on the test set of the trained network was compared with that of the current optimised network. The optimised network among the 100 networks was obtained after 100 iterations of the training-testing process. The procedural steps of the proposed method are outlined below:

Algorithm: Iterative training-testing procedure
Input: optimised 2-layer MLP, merged_data, K (iterations)
Output: optimised iterative network

```
1.   net ← optimised 2-layer MLP;
2.   optimised_iter_net ← net;
3.   t ← 1;
4.  for (t ≤ K) {
5.          Randomly split merged_data into balanced training set,
            validation set, test set;
6.          Set the initial weights of the training algorithm to
            the weights of net;
7.          net ← train(net,train_set,valid_set);
8.          mean_rmse ← test(net,test_set);
9.          If mean_rmse < mean_rmse of optimised_iter_net
10.         Then    optimised_iter_net ← net;
11.         t ← t + 1;
12.}
13. Output the optimised_iter_net;
```

3.5 Results

3.5.1 The Optimised 1-Hidden Layer MLP

The optimised 1-hidden layer MLP among the 30 MLPs trained has a mean RMSE of 2.23 and consists of 18 hidden nodes (Fig. 3.3). The RMSE of each break size of the optimised network on the test set is illustrated in Table 3.1 and Fig. 3.4.

3.5.2 The Optimised 2-Hidden Layer MLP

The optimised 2-hidden layer MLP among the 30 2-hidden layer MLPs trained has a mean RMSE of 1.59 and consists of 12 nodes in each hidden layer and 600 weights (Fig. 3.5). The optimised 2-hidden layer MLP has a better performance

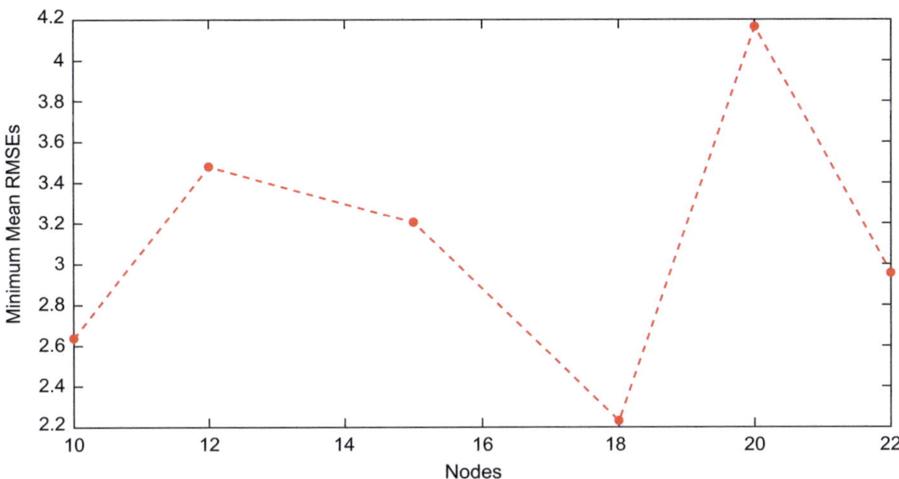

Fig. 3.3 Minimum mean RMSEs among five training times of the six 1-hidden layer MLP architectures

Table 3.1 The RMSEs of each break size of the optimised 1-hidden layer MLP

0%	20%	60%	100%	120%	200%	Mean RMSE	Standard deviation of RMSEs
0.06	1.39	2.01	3.37	3.38	3.16	2.23	1.34

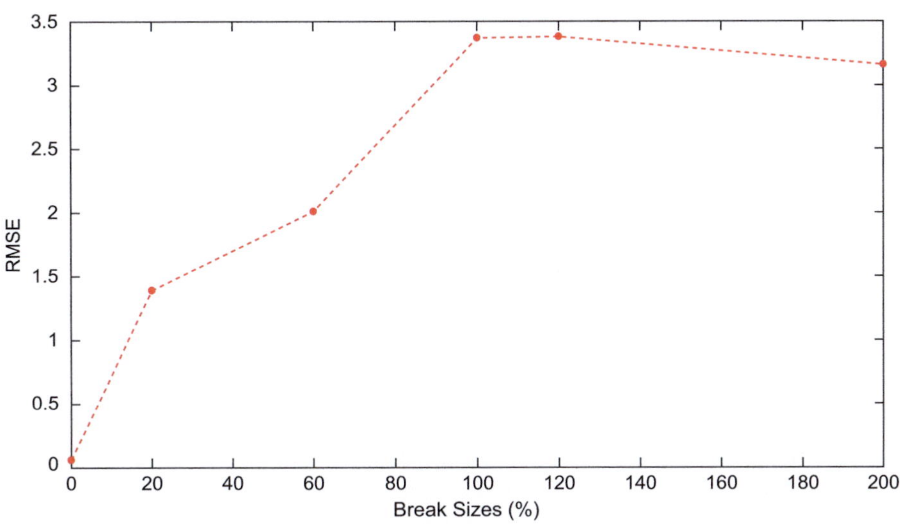

Fig. 3.4 RMSE of each break size of the optimised 1-hidden layer MLP

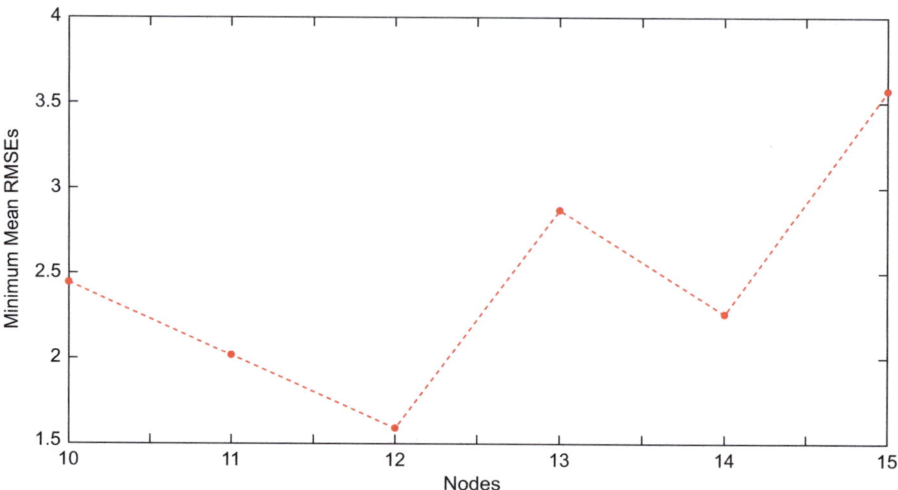

Fig. 3.5 Minimum mean RMSEs among five training times of the six 2-hidden layer MLP architectures

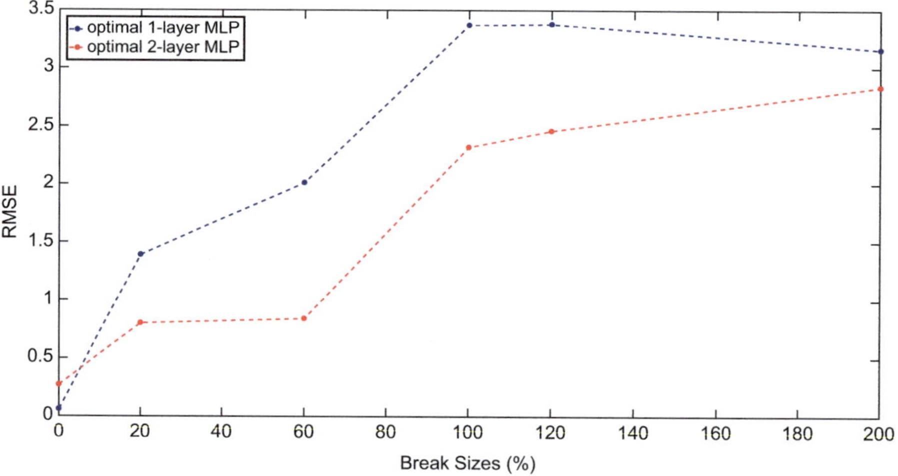

Fig. 3.6 Performance comparison of the optimised 2-layer MLP and the optimised 1-layer MLP

(mean RMSE 1.59) and a more stable performance (standard deviation 1.075) than the optimised 1-layer MLP in detecting break sizes. The optimised 2-layer MLP has significantly smaller RMSEs than the optimised 1-layer MLP on all break sizes except break size 0% (Fig. 3.6 and Table 3.2).

Table 3.2 The RMSE of each break size of the optimised 2-hidden layer networks

0%	20%	60%	100%	120%	200%	Mean of RMSEs	Standard deviation of RMSEs
0.27	0.80	0.84	2.32	2.46	2.84	1.59	1.075

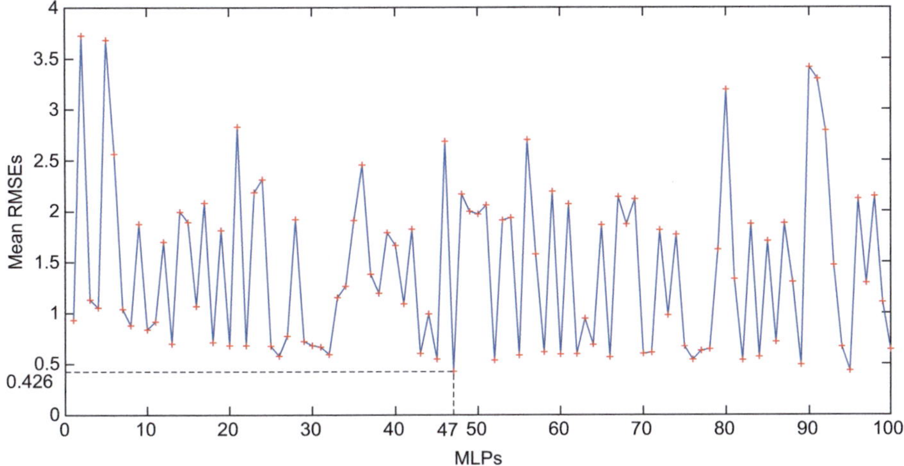

Fig. 3.7 Mean RMSEs of the 100 MLPs

3.5.3   *Training the Optimised 2-Hidden Layer MLP on Linear Interpolation Dataset and Transient Dataset*

The optimised 2-layer MLP was trained and tested iteratively 100 times on the merged dataset to obtain a MLP with better performance than the optimised 2-layer MLP. The mean RMSEs of the 100 networks are compared in Fig. 3.7. The mean RMSE of the 47th network is the smallest (0.4261). The standard deviation of the RMSEs of the 47th network on all the break sizes is 0.2342. Therefore, the 47th network is the optimised network of the iterative training-testing process. The mean RMSE of the 95th network is 0.434 and the 2nd smallest. The RMSE of each break size of the 47th network is smaller than that of the optimised 2-hidden layer MLP (Fig. 3.8 and Table 3.3). The standard deviation of the RMSEs of the 47th MLP is 0.2342 which is smaller than that of the optimised 2-hidden layer MLP. Therefore, the 47th MLP has a significantly more stable performance (less variation of performance) than the optimised 2-layer MLP in detecting different break sizes. Therefore, the performance of the 47th network is much higher than that of optimised 2-layer MLP.

Comparison of the outputs of the optimised network and the break size targets on the test set is illustrated in Fig. 3.9. For all the 6 break sizes 0%, 20%, 60%, 100%, 120% and 200%, most of the optimised network outputs are identical to the targets. For each break size, the mean of the optimised network outputs is very similar to the target break size (Table 3.4). The variations of the networks outputs are small.

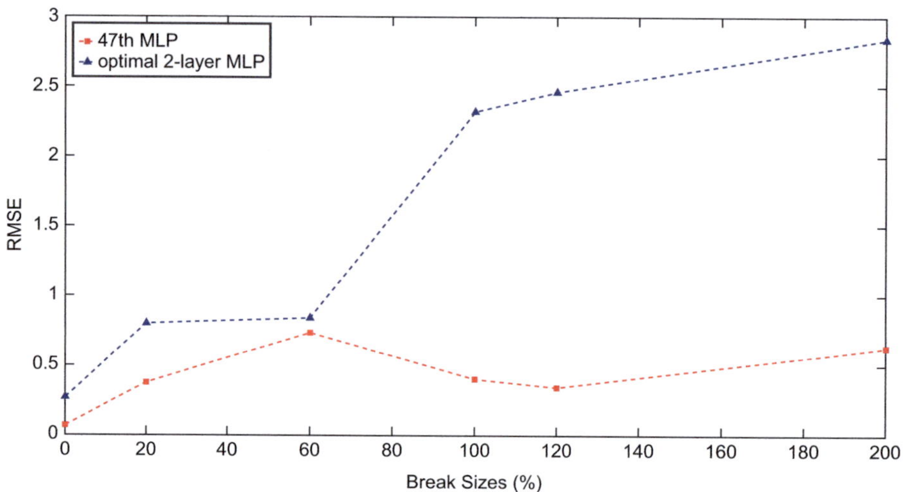

Fig. 3.8 Performance comparison of the 47th MLP and the optimised 2-layer MLP

Table 3.3 The RMSE of each break size of the 47th MLP

0%	20%	60%	100%	120%	200%	Mean of RMSEs	Standard deviation of RMSEs
0.0690	0.3761	0.7344	0.4030	0.3436	0.6307	0.4261	0.2342

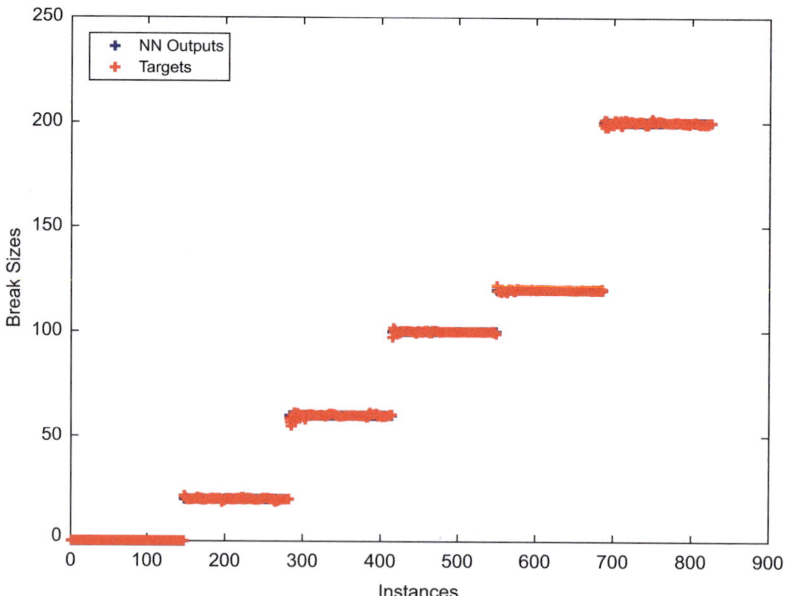

Fig. 3.9 Comparison of the outputs of the 47th network and the target break sizes of test set

Table 3.4 The mean and standard deviation of the optimised network outputs corresponding to each break size

Target break sizes	0%	20%	60%	100%	120%	200%
Mean of network outputs (%)	0.07	20.02	59.99	99.96	120.00	200.03
Standard deviation of network outputs	0	0.37	0.74	0.40	0.34	0.63

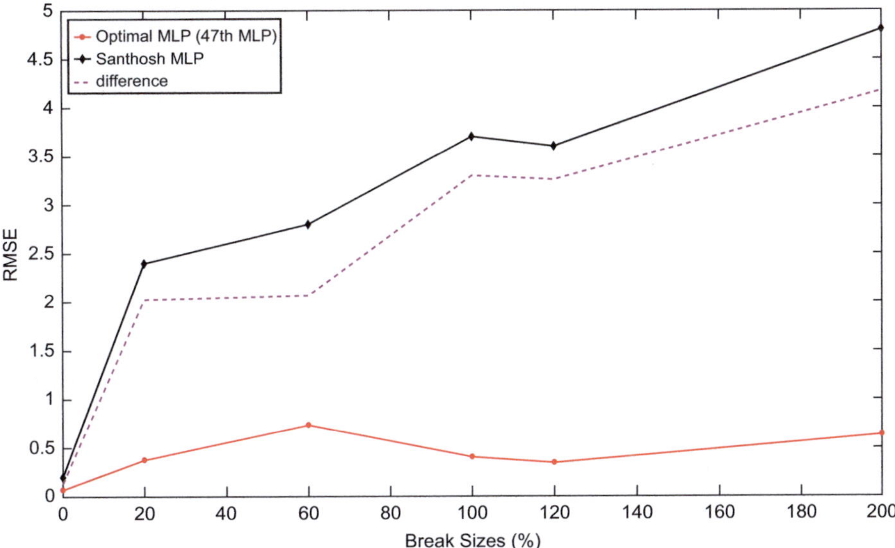

Fig. 3.10 Performance comparison of the optimised MLP and the Santhosh et al.'s network [5]

3.5.4 *Performance Comparison with the Neural Network of the Previous Work*

The Santhosh et al.'s MLP [5] has 37 inputs, 2 hidden layers and 3 output nodes which output break size, location of the breaks and the availability of the ECCS. In contrast, this work focuses on detecting the beak size of an IH of the PHT, and the optimised MLP (the 47th MLP) detects a break size of an IH rather than the location of the break and the availability of ECCS. The performance of the optimised MLP is compared with the performance of Santhosh et al.'s MLP with regard to break size detection (Fig. 3.10). The optimised MLP has smaller RMSEs than the Santhosh et al.'s MLP [5] on all the 6 break sizes with the largest difference in RMSE being 4.1693 at break size 200% and the smallest difference being 0.131 at break size 0% (Fig. 3.10). The mean RMSE (0.4261) of the optimised MLP is smaller than that of the Santhosh et al.'s MLP (2.9167). The optimised MLP has a significantly more stable performance (standard deviation 0.2342) than the Santhosh's MLP (standard deviation 1.5677) in detecting the break sizes. However, it may be noted that the RMSE in Santhosh et al.'s MLP [5] has been computed based on three outputs: break size, the break location and the status of ECCS.

3.5.5 Performance Comparison with Exhaustive Training of All 2-Hidden Layer Architectures

All the 2-hidden layer MLP architectures with each hidden layer consisting of 5–40 hidden nodes were trained on the training set and tested on the test set. The validation set was used to validate the performance of each architecture during training. Each architecture was trained five times. The optimised MLP among the 6480 ($36\times36\times5$) MLPs has 37 inputs, 30 nodes in the first hidden layer and 18 nodes in the second hidden layer. The training time to find the optimised MLP is approximately 25 h 35 min on a Windows 10 desktop computer with a Core-i7 CPU of 3.6 GHz and a 16 GB RAM. Thereafter, the optimised MLP was iteratively trained 100 times on the linear interpolation dataset and the transient dataset. The optimised MLP of the exhaustive training was chosen from the 100 trained MLPs. The training time of training the 100 MLPs is approximately 50 min. The performance of the optimised MLP of this work is compared with the performance of the optimised MLP of exhaustive training (Fig. 3.11 and Table 3.5). Although the mean RMSE (0.4261) of the optimised MLP of this work is larger than that of the optimised MLP of exhaustive training (0.2751), the difference in mean RMSE is small (0.151). The

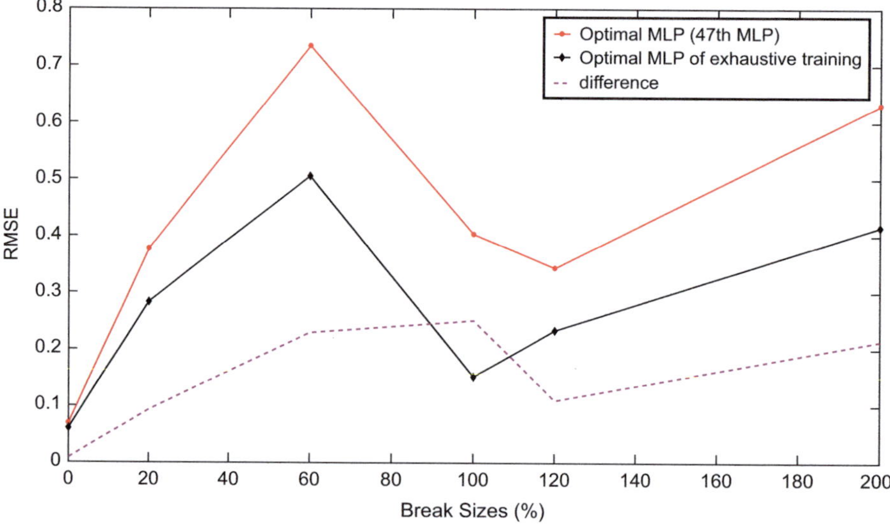

Fig. 3.11 Performance comparison of the optimised MLP (47th MLP) and the optimised MLP of exhaustive training

Table 3.5 The RMSE of each break size of the optimised MLP of exhaustive training

0%	20%	60%	100%	120%	200%	Mean of RMSEs	Standard deviation of RMSEs
0.0604	0.2831	0.5051	0.1521	0.2337	0.4157	0.2751	0.1647

RMSE of each break size of the optimised MLP of exhaustive training is smaller than that of the optimised MLP of this work (Fig. 3.11). However, the difference in RMSE is small with the minimum difference of 0.0086 at break size 0% and the maximum difference of 0.2509 at break size 100% (Fig. 3.11). Although the standard deviation (0.1647) of the optimised MLP of our exhaustive training is smaller than that (0.2342) of the optimised MLP of this work, the difference in standard deviation is very small (0.0695).

3.6 Discussion

The proposed methodology trained 60 MLPs (30 1-hidden layer MLPs and 30 2-hidden layer MLPs) on the transient dataset and trained 100 2-hidden layer MLPs on the transient dataset added with the linear interpolation dataset to find an optimised 2-hidden layer MLP. The training time is approximately 70 min on the Windows 10 desktop computer with a Core-i7 CPU of 3.6GHz and a 16 GB RAM. Therefore, the main advantage of the proposed methodology is that it finds a neural network with very high performance at a much faster speed than exhaustive training of all 2-hidden layer architectures (training time: 26 h 25 min approximately). This is due to the following key steps of the proposed methodology:

1. The architectures of 2-hidden layer MLPs are determined based on the number of the weights of the optimised 1-hidden layer MLP found and trains each architecture 5 times to find an optimised 2-hidden layer MLP. This step significantly reduces the training time by only training a number of 2-hidden layer balanced architectures.
2. The optimised 2-hidden layer MLP is iteratively trained on the linear interpolation dataset and the transient dataset. This step further improves the generalisation performance of the optimised 2-hidden layer MLP.The optimised MLP of exhaustive training (48 hidden nodes) has twice as many hidden nodes as the optimised MLP of this work (24 hidden nodes). However, the difference in performances between the 2 optimised MLPs is small. Therefore, the optimised 2-hidden layer MLP of this work tends to have better generalisation performance than the optimised MLP of exhaustive training.

3.7 Conclusion

This work has proposed an efficient methodology to design a 2-hidden layer MLP of high performance for detection of loss of coolant accident in a PHWR. The performance of the proposed methodology is outstanding in detecting break sizes of IHs of a PHWR. We have tackled this predictive data analytics challenge using well-designed neural networks. Our neural network architectures design methodology aimed at tackling the challenging nature of transient datasets which exhibit big data characteristics. Feature selection algorithms [25, 26, 28] select the most relevant features of a dataset which can be used to build predictive models with better generalisation performances than the models built using the set of all the features of the dataset. Feature selection can be applied to the training set to obtain networks with better generalisation performance than that of the optimised 2-hidden layer MLP. Transient datasets representing the locations of breaks and the availabilities of ECCS can be generated using RELAP-3D. Then, different classifiers such as neural networks, Bayesian networks and support vector machines (SVMs) can be trained on the transient datasets to detect the locations of breaks and the availabilities of ECCS.

Acknowledgements The authors would like to thank EPSRC for their financial support under the grant number of EP/M018717/1. The Engineering and Physical Sciences Research Council (EPSRC) is the UK's main agency for funding research in engineering and the physical sciences.

References

1. Maio, F., et al. (2017). Safety margin sensitivity analysis for model selection in nuclear power plant probabilistic safety assessment. *Reliability Engineering and System Safety, 162*, 122–138.
2. Safety margins of operating reactors: Analysis of uncertainties and implication for decision making. (2003). Technical report IAEA-TECDOC-1332, International Atomic Energy Agency (IAEA). http://www-pub.iaea.org/MTCD/publications/PDF/te_1332_web.pdf
3. Procedures for Conducting Probabilistic Safety Assessments of Nuclear Power Plants (Level 1). (1992). Safety series, IAEA https://gnssn.iaea.org/Superseded%20Safety%20Standards/Safety_Series_050-P-4_1992.pdf
4. Na, M. G., et al. (2004). Estimation of break location and size for loss of coolant accidents using neural networks. *Nuclear Engineering and Design, 232*, 289–300.
5. Santhosh, T. V., et al. (2011). A diagnostic system for identifying accident conditions in a nuclear reactor. *Nuclear Engineering and Design, 241*, 177–184.
6. Barlett, E. B., & Uhrig, R. E. (1992). Nuclear power plant status diagnostics using an artificial neural network. *Nuclear Technology, 97*, 272–281.
7. Guo, Z., & Uhrig, R. E. (1992). Use of artificial neural networks to analyse nuclear power plant performance. *Nuclear Technology, 99*, 36–42.
8. The RELAP5-3D Code Development Team. (2014). RELAP5-3D code manual volume V: User's guidelines, INL-EXT-98-00834, Revision 4.2, Idaho National Laboratory, USA
9. Hazewinkel, M. (2001). Linear interpolation. In M. Hazewinkel (Ed.), *Encyclopedia of mathematics*. Dordrecht: Springer.
10. Le, H. V. (2002). Large LOCA analysis of Indian Pressurized Heavy Water Reactor – 220 MWe. *Nuclear Science and Technology, 1*, 12–17.
11. Modular Accident Analysis Program 5 (MAAP5) Applications Guidance: Desktop Reference for Using MAAP5 Software – Phase 1 Report (2014). EPRI, Palo Alto, CA: 3002003113.
12. Volkanovski, A., et al. (2007). *An application of the fault tree analysis for the power system reliability estimation*. International Conference Nuclear Energy for New Europe, Slovenia
13. Karanki, D., et al. (2015). A dynamic event tree informed approach to probabilistic accident sequence modelling: Dynamics and variabilities in medium LOCA. *Reliability Engineering and System Safety, 142*, 78–91.

14. Karanki, D. et al. (2011). *Discrete dynamic event tree analysis of MLOCA using ads-trace.* International Topical Meeting on Probabilistic Safety Assessment and Analysis 2011, PSA 2011, 1:pp. 610–622
15. Buckingham, R., & Graham, A. (2012). Nuclear Snake-arm Robots. *Industrial Robot: An International Journal, 39*(1), 6–11.
16. Ferguson, T. A., & Lu, L. (2017). Fault tree analysis for an inspection robot in a nuclear power plant. *IOP Conference Series: Materials Science and Engineering, 235.*
17. Zio, E., Maio, F. D., & Stasi, M. (2010). A data-driven approach for predicting failure scenarios in nuclear systems. *Annals of Nuclear Energy, 37*(4), 482–491.
18. Baraldi, P., Razavi-Far, R., & Zio, E. (2011). Bagged ensemble of Fuzzy C-Means classifiers for nuclear transient identification. *Annals of Nuclear Energy, 38*(5), 1161–1171.
19. Wang, W. Q., Golnaraghi, F. M., & Ismail, F. (2004). Prognosis of machine health condition using neuro-fuzzy systems. *Mechanical Systems and Signal Processing, 18*(4), 813–831.
20. Wei, X., Wan, J., & Zhao, F. (2016). Prediction study on PCI failure of reactor fuel based on a radial basis function neural network. *Science and Technology of Nuclear Installations, 2016,* 1–6.
21. Souza, T. J., Medeiros, J. A., & Gonçalves, A. C. (2017). Identification model of an accidental drop of a control rod in PWR reactors using thermocouple readings and radial basis function neural networks. *Annals of Nuclear Energy, 103,* 204–211.
22. Secchi, P., Zio, E., & Maio, D. F. (2008). Quantifying uncertainties in the estimation of safety parameters by using bootstrapped artificial neural networks. *Annals of Nuclear Energy, 35*(12), 2338–2350.
23. Back, J., et al. (2017). Prediction and uncertainty analysis of power peaking factor by cascaded fuzzy neural networks. *Annuals of Nuclear Energy, 110,* 989–994.
24. Guimarães, A., & Lapa, C. (2007). Adaptive fuzzy system for fuel rod cladding failure in nuclear power plant. *Annuals of Nuclear Energy, 34,* 233–240.
25. Han, J., Kamber, M., & Pei, J. (2011). *Data Mining: Concepts and Techniques (3rd edition).* San Francisco: Morgan Kaufmann Publishers.
26. Hand, D. J., Smyth, P., & Mannila, H. (2001). *Principles of data mining.* Cambridge, MA: MIT Press.
27. He, H., & Garcia, E. A. (2009). Learning from imbalanced data. *IEEE Transactions on knowledge and data engineering, 21*(9), 1263–1284.
28. Witten IH, Frank E (2005) Data mining: Practical machine learning tools and techniques, 2nd. Morgan Kaufmann Publishers Inc., San Francisco
29. Bishop, C. M. (1995). *Neural networks for pattern recognition.* New York: Oxford university press.
30. Bishop, C. M. (2006). *Pattern recognition and machine learning.* Singapore: Springer.

Chapter 4
Evolutionary Deployment and Hill Climbing-Based Movements of Multi-UAV Networks in Disaster Scenarios

D. G. Reina, T. Camp, A. Munjal, S. L. Toral, and H. Tawfik

4.1   Introduction

Every year, millions of people are affected by natural and man-made disasters involving large expanses of land. Such disasters include earthquakes, tsunamis, volcano eruptions, hurricanes, tornados, floods, and terrorist attacks. Governments all around the world spend huge amounts of resources not only on preparation for such events but also on reconstruction in their aftermath. These traumatic events can severely damage both public and private infrastructure and can dramatically compromise people's welfare. Studies suggest that the first 72 hours post-disaster are extremely important [1, 2]. This period is called the "golden relief time" [1]. After the golden relief time, the probability of finding survivors is very low. Consequently, coordination of first responders and victims is of paramount importance. Communications, both in general and among first responders, are vitally important to efficiently coordinate rescue efforts during this critical window.

D. G. Reina (✉)
Engineering Department, Loyola Andalucía University, Seville, Spain
e-mail: dgutierrez@uloyola.es

T. Camp
Computer Science Division, Colorado School of Mines, Golden, CO, USA

A. Munjal
Testplant Inc., Boulder, CO, USA

S. L. Toral
Electronic Engineering Department, University of Seville, Seville, Spain

H. Tawfik
School of Computing, Creative Technology and Engineering, Leeds Beckett University, Leeds, UK
e-mail: h.tawfik@leedsbeckett.ac.uk

© Springer International Publishing AG, part of Springer Nature 2018
M. M. Alani et al. (eds.), *Applications of Big Data Analytics*,
https://doi.org/10.1007/978-3-319-76472-6_4

A thorough survey of interviews with German first responders indicates that the first few minutes of an emergency are the most important [3]. Furthermore, while a basic communication infrastructure has to be established, rescuers should not spend their valuable time in this effort.

Nowadays, people commonly communicate with each other using their cell phones, i.e., smartphones with Internet access provided either by their telecommunication operator or by connecting to wireless fidelity (Wi-Fi) access points (APs). Chat applications like WhatsApp and Google Talk, or social networks like Facebook and Twitter, have changed the way that people communicate. It should be noted, however, that the use of the above-mentioned Internet-based applications could be compromised by damage to the communication infrastructure, leaving many people isolated and unable to communicate. Moreover, even traditional communication services, like voice calls and text messages, will not be possible in the event of major damage. Device-to-device communications, like the ones established by using Wi-Fi direct, are limited due to the slow market penetration of such technology so far.

Therefore, alternative ad hoc communication and IoT-based (Internet of Things) infrastructures should be deployed in a rapid and self-configured manner to allow interpersonal communication and access to the Internet [4]. In this chapter, we study intelligent deployment and tactical movement of mobile APs that act as 0th responders, arriving at the disaster area as soon as possible to provide communication services. In our work, these APs are drones or Unmanned Aerial Vehicles (UAVs) equipped with Wi-Fi transceivers that can move throughout the disaster area. The most optimal deployment of the 0th responders will depend on several factors. First, deployment depends on the available information regarding the disaster scenario, in that there exists a need to collect certain information before the arrival of the 0th responders at the disaster site. Such information can be collected via satellite images from different sources, such as people living near the disaster area and satellite images, among others. If the requisite information is available, we can design an initial deployment of the 0th responders to cover the most important target points or areas. After that, the drones' positions should adapt to the conditions of the disaster, using local information collected directly from the disaster area. Following the initial deployment of the 0th responders, the ultimate objective is to find possible victims that have not been found during the deployment. For the first deployment problem, we propose an evolutionary algorithm, i.e., a genetic algorithm (GA). Then, we adopt a local search such as the hill climbing algorithm (HCA) to explore new areas and adapt the previous deployment to the real conditions of the disaster site.

One important feature of the proposed deployment and subsequent tactical movements is that drones should form a connected UAV network, in that there are no isolated drones and that every drone is reachable from every single other drone. This connectivity requirement exists because one of our main objectives is to provide Internet services to victims. To make this possible, the UAV network should use another long-range communication technology (i.e., satellite communications). The idea is that one of the drones will be equipped with a satellite transceiver and it will share the Internet connection with the rest of the drones forming the network. Thus, in an unconnected mesh network, some drones would not be able to provide Internet services to victims in their locations. It is worth highlighting that the ad hoc term refers to the idea that the topology of the network is dynamic, and those nodes, in this case drones, are responsible for routing the data packets in the network. In fact, we note that wireless multi-hop ad hoc networks [5] have already been envisioned as an attractive communication paradigm to be utilized in disaster scenarios [5]. However, this study differs from the direct application of the ad hoc paradigm in disaster scenarios [6] in that we aim to deploy an alternative and dynamic communication infrastructure to which the victims can easily connect with their cellular phones in the same way that they connect to Wi-Fi spots at home or in public spaces.

Typically, victims and first responders will access the network using common portable devices such as smartphones, tablets, etc. Thus, unlike the specialized mobile radios normally used by first responders, our proposed network is easy to access and provides widespread usability. Consider, for example, the TETRA (terrestrial trunked radio) technology that is widely used in Europe [7]. TETRA terminals allow first responders to establish wireless communications; however, they lack interoperability with other ubiquitous free band wireless technologies like Wi-Fi and Bluetooth. This shortcoming greatly hinders cooperation between victims and first responders since victims are not equipped with TETRA terminals. Consequently, in a disaster scenario, many people can remain isolated unless an alternative communication infrastructure is rapidly deployed. The preexisting cellular-based communication infrastructure can take several days or weeks to repair, which is well outside the "golden relief time."

This work attempts to tackle an important disaster management application in this IoT and big data era. The main aim is to determine optimum or near optimum solutions in a potentially very large and complex search space. This is due to high dimensionality and huge increase of parameters and combinatorics, with the increase in the number of UAVs, and size and resolution of the disaster terrain. Therefore, we consider this an application of data analytics, namely, decision analytics, problem to address using computational intelligence techniques.

The chapter continues as follows. Section 4.2 includes some relevant related work on two similar topics, such as the deployment problem and the mobility models for disaster scenarios. Section 4.3 describes the main features of disaster scenarios including scenario layout, mobility of victims, mobility of 0th responders, and communications. Section 4.4 presents the proposed approach and outlines the evolutionary and the local search algorithms used in our work. Section 4.5 presents and analyzes our simulation results. Finally, Section 4.6 concludes this chapter.

4.2 Related Work

We divide the related work into two subsections. Section 4.2.1 reviews related works that address the deployment problem in disaster scenarios, while Sect. 4.2.2 is devoted to reviewing prior research on mobility models for disaster scenarios.

4.2.1 Deployment Problem

A thorough survey on the application of evolutionary algorithms in disaster scenarios can be found in [8]. Among the problems described in [8], the ones related to location are most pertinent to this chapter. The location problems in disaster scenarios are focused on finding the best positions for fire stations, medical services, shelters, etc. These optimization problems are based on the study and analysis of the topography of a disaster area. The mentioned problems are related to the work presented in this chapter, e.g., the deployed facilities should cover many victims in a disaster scenario.

In [9], the authors propose the use of UAVs to deploy 5G dynamic cellular-based stations in the aftermath of a disaster scenario. They use a brute force algorithm to find the most optimal positions for the drones. A brute force algorithm tries to evaluate all possible solution of a given problem. However, this type of strategy is not suitable for NP-hard problems like the coverage problem presented in this chapter.

In [10], the authors use genetic programming for search tasks of multiple UAVs. A team of UAVs is tasked with exhaustively covering a predefined search area, which is divided into target beacons, and then returning to a base location. This work is more focused on a military scenario, e.g., they consider that the drones can be destroyed due to hostile situations. The possible movements of a drone are defined as a decision tree, and genetic programming is used to determine the best moving strategy.

In [11], the authors use a multi-population genetic algorithm to solve a multi-objective coverage problem. The coverage problem consists of weighting three important features of an UAV network, such as number of ground nodes covered, fault tolerance, and accessibility of the network. The authors select the weights of the three mentioned objectives so that the number of covered ground nodes has the highest importance and the accessibility the lowest. To solve such multi-objective coverage problem, the authors proposed a genetic algorithm that exploits multiple subpopulations evolving in parallel with different layouts in terms of genetic operators. The subpopulation exchange solutions through a migration scheme. The authors compare the proposed genetic algorithm with classical ones, demonstrating that their approach outperformances classical genetic algorithms. This work is a step forward since once the drones are placed in the incident site, they will try to explore new areas to find uncovered ground nodes.

In [12], the authors propose HMADSO, a bio-inspired algorithm for cooperative rendezvous and task allocation in flying ad hoc networks (FANETs). The authors consider the biological communication and self-organized abilities of two birds, such as hill myna and desert sparrow to optimize both the mentioned tasks in FANETs. The HMADSO algorithm allows nodes to cluster themselves according to topological features. The proposed approach outperforms other bio-inspired optimization algorithms, such as ant colony optimization (ACO) and bee colony optimization (BCO).

In [13], the drones' deployment and movements are self-organized. Several nature-inspired optimization algorithms are used to maximize the number of ground nodes (victims and first responders) under coverage, i.e., the serviced nodes. UAVs share between each other the identity of the ground nodes serviced by each one, which is considered a set. A metric that measures the dissimilarity between two sets can be used to evaluate the victims shared between different UAVs, i.e., those victims that are under the coverage area of several UAVs. The selected metric is the Jaccard distance, which may take values within the range [0,1]. The target Jaccard is calculated by different optimization algorithms such as hill climbing and simulated annealing. For each iteration, the algorithms return the value that maximizes the number of serviced ground nodes. By doing this, the UAV network adapts the UAV positions according to the ground nodes movements. The algorithms also penalize the solutions that disconnect the UAVs from the network; thus, a connected network is guaranteed.

The deployment problem of Wi-Fi wireless routers, using evolutionary algorithms, has already been studied in several works [14–16]. Our study, however, differs from the previous work for several reasons. First, the previous work does not consider that the Wi-Fi routers have to form a connected mesh network. Second, they assume that the Wi-Fi routers have to be placed in a grid, so the discrete search space is reduced in comparison to the continuous search space considered in this study. Third, in some studies, including [15, 16], the authors do not limit the number of clients that a Wi-Fi router can serve simultaneously.

In [17], the authors propose a flocking-based approach to improve the performance of a MANET. They divide the nodes of the network into two categories: users and agents. The objective of the agents is to be placed in suitable positions to improve the communication among users. The flocking algorithm relies on the following three features: (a) cohesion (attempt to stay close to nearby flock mates), (b) separation (avoid collisions with nearby flock mates), and (c) alignment (attempt to match velocity with nearby flock mates). They conduct several experiments to validate their approach. A similar work is presented in [18], but, in this case, the authors use a multi-objective genetic algorithm to optimize the performance of the network in terms of four output parameters: maximizing communication coverage, minimizing the active structures' costs, maximizing the total capacity bandwidth, and minimizing the noise level in the network.

In [19], the connectivity of crew members acting in a disaster scenario is improved by the deployment of static auxiliary beacon nodes that are used as packet forwarders. The optimization problem is solved by applying a single objective genetic algorithm. In [19], the connectivity of the network is measured as the reachability achieved by broadcasting packets sent by the crew members. The main issue of the solution proposed in [19] is that the authors find the optimal positions of the auxiliary beacon offline. In addition, they only consider the connectivity among crew members. Consequently, the victims are not modeled in [19].

In [20], an algorithm is proposed to solve the convex optimization problem of moving several robots toward a target point while maintaining the connectivity of the robots. Consequently, the wireless robots need to get closer to the destination but without losing the wireless connectivity that exists. This is a reconfiguration problem, where robots should reconfigure themselves to reach the target destination. The authors propose both a centralized and distributed version of their algorithm. In [21], the authors present an algorithm to deploy wireless robots to connect a mobile user to a base station through a multi-hop communication path. The authors calculate the minimum number of wireless robots necessary to guarantee the shortest path between the mobile user and the base station. Their approach is based on equidistant separation of robots along the shortest path. In addition, the authors evaluate their approach considering obstacles in the communications among nodes.

Our work stands out from previous work in this scientific area for several reasons. First, we consider that the information on the target scenario is limited. Second, the deployment of drones should guarantee that the drones form a connected mesh network. Third, we limit the number of clients that can be served by the same Wi-Fi router. Fourth, our proposed approach is not aimed at improving the performance of an ad hoc network; it is intended to improve the communication among the victims and the 0th responders. Fifth, the drones do not know the position of all victims during the adaptation to the real conditions phase; the drones only know a percentage of the victims' positions in the initial deployment phase, which is denoted by K (described in more details later). Consequently, the problem presented in this chapter is an exploration/search problem significantly different from the reconfiguration problem presented in [20].

4.2.2 Mobility Models for Disaster Scenarios

The mobility of first responders has already been modeled in [22], where the authors present the disaster area mobility model. The Disaster Area mobility model is based on a method called "separation of the room" [22, 23]. In the Disaster Area mobility model, a disaster scenario is divided into different context-based areas. These areas are incident site, casualty treatment area, transport zone, and technical operation command zone. One desired feature that does not exist in the Disaster Area mobility model proposed in [22] is the mobility of possible victims in the disaster area. In [24], the authors present CORPs, a synthetic mobility model for first responders. In this mobility model, attraction points, called attention events, are defined such that first responders move to these points. Unfortunately, the authors of [22] also do not consider the mobility of possible victims.

In [25], the authors present the human behavior for disaster areas (HBDA) mobility model. It simulates the behavior of rescue teams performing search-for-victim operations. This behavior is represented by a set of algorithms that models a group of people scouting unexplored areas; the scouting people are distributed over the scenario while maintaining a line of sight with other members involved in the search operation.

Another possibility is to use scenarios based on real maps and integrate the mobility of nodes restricted to those maps in a network simulator. This possibility is the basis for map-based mobility models. For example, in [26], the authors use a map-based mobility model with a map of the city of Loja in Spain. In [27], the authors propose a map-based mobility model for a delay-tolerant network (DTN) that considers the mobility of both rescue workers and victims. The victims move toward evacuation centers, whereas the rescue workers move toward the victims to supply relief food. With this map-based mobility model, we achieve a more accurate model of the disaster scenario; however, the mobility of nodes is still synthetically generated.

This chapter is not strictly about the design of a mobility model for drones in disaster scenarios. Our adaptation to the real conditions, however, means finding the optimal tactical movements of drones at the disaster scenario, which is a similar problem. Regarding the previous work in this field, this study takes into consideration both the victims and the crew members. This feature is important since the movements of a rescue team should always depend on the positions and movements of the victims in the disaster scenario. The proposed approach also differs from previous path planning projects that are focused on finding optimal trajectories of drones in different target application areas [28]. Path planning of intelligent vehicles is also an active research topic for the application of global search algorithms, e.g., evolutionary algorithms [29, 30].

4.3 Modeling Disaster Scenarios

Modeling disaster scenarios is a challenging task due to the high number of parameters and variables to be considered in a possible model. In [3], the author indicates that a classification of disaster scenarios is impossible because of the high unpredictability and variability among different disaster scenarios. In general, the origin of a disaster can vary from one scenario to another. This fact makes it even more difficult to model unpredictable and dynamic features, such as mobility of victims and/or rescue teams, density of victims involved, existing communication infrastructure that might be still functioning, terrain conditions, aspects of urban environments (e.g., buildings and roads) that may be destroyed or modified by the disaster, etc.

We divide the disaster area features into four main categories: (1) disaster scenario layout, (2) mobility of victims, (3) rescue teams (0th responders and 1st responders), and (4) communications. Figure 4.1 illustrates these four main components and their interactions to be considered for modeling disaster scenarios. In the following four subsections, we provide more details about each category.

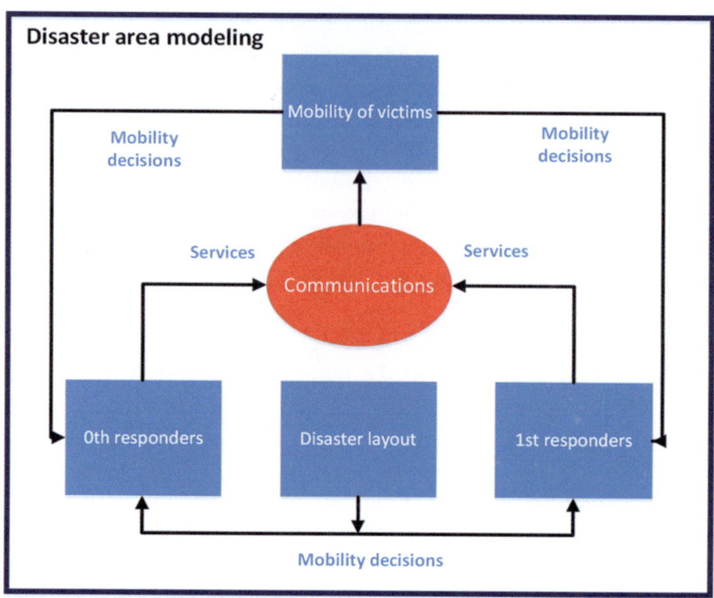

Fig. 4.1 Disaster area modeling

4.3.1 Disaster Scenario Layout

Based on the disaster scenario layout, we can categorize disaster scenarios as either urban or rural. In the former, the disaster occurs in a city or town causing damage to both the urban infrastructure (e.g., building and roads) and to citizens. Depending on the nature of the damage, the disaster can be localized to a part of the city, such as the collapse of the World Trade Center of New York in 2001 or the disaster that took place on March 11, 2004, in the city of Madrid, or it can affect a much larger area, such as the tsunami that hit northeastern Japan in 2011.

On the other hand, in rural scenarios, the disaster mostly affects people living in the surrounding area or involved in the accident and, of course, the natural environment. Several examples of rural disaster scenarios are forest fires, floods, and, more recently, the Germanwings Flight 9525 that was intentionally crashed into the French Alps in March 2015 causing 150 fatalities.

4.3.2 Mobility of Victims

The mobility of victims will strongly depend on the disaster scenario layout. In urban scenarios, one expects to find trapped victims in their homes or cars. Victims in outside locations typically try to find shelter as quickly as possible by running or using their vehicles; however, roads can be partially destroyed, which decreases the mobility of victims and vehicles. Consequently, the mobility of victims, in most cases, can be limited to within a certain urban area. In rural scenarios, the area of mobility can be unknown for victims who try to facilitate rescue by moving to a better location. Of course, victims with severe injuries will remain static in the incident location waiting for help. Another important aspect regarding the mobility of victims is the tendency to form groups. In harsh situations, people feel more comfortable and protected if they remain together.

4.3.3 0th Responders

In our context, 0th responders are UAVs or drones that can quickly deploy in a disaster area before (or as) first responders arrive. These 0th responders perform in a coordinated manner, communicating with each other wirelessly.

The use of UAVs as 0th responders enables the deployment of the two main communication networks defined in [28] for disaster relief operations: disaster recovery networks (DRNs) and search and rescue networks (SRNs). The objective of a DRN is to provide emergency support to victims and crew members taking part in rescue operations. To accomplish this task, the 0th responders need to be strategically placed in the disaster area to provide the maximum communication coverage possible to victims. The main goal of the SRN is to find and track victims. To this end, 0th responders will explore new locations in the disaster areas to locate new victims.

According to the goals defined by DRNs and SRNs, we divide the movements of 0th responders into two modes:

- Coverage mode: the objective is to place drones in the disaster scenario to provide the maximum communications coverage possible to victims.
- Exploration mode: the objective is to have the drones explore new areas to search for other possible victims.

4.3.4 Communications in Disaster Scenarios

From previous disaster scenarios research [5, 22], it is reasonable to assume that disasters can disable or destroy preexisting communications infrastructure. Moreover, disasters can occur in isolated areas, such as mountains, where cellular-based communications are limited by coverage issues. Consequently, an alternative communication network is needed for victims to communicate with rescue teams via hand-portable devices such as smartphones. Multi-hop ad hoc networks have been envisioned as an appealing technology for disaster scenarios [5]. Under the ad hoc paradigm, people using electronic wireless devices can communicate with each other without a communication infrastructure. Multi-hop ad hoc networks allow two types of communications, broadcasting one-to-all communications [31, 32] and unicast communications via routing protocols [33].

In our proposed communication infrastructure, the 0th responders form a multi-UAV network using ad hoc communications. Furthermore, the 0th responders function as access points for victims. It is important to recall that victims will not connect to the mesh network using ad hoc mode. The victims will access the mesh network using normal mode, similar to how people normally connect to Wi-Fi. This feature is important because the ad hoc mode in the Wi-Fi transceivers included in smartphones is limited. For example, in Android phones, the ad hoc mode only activates if the phone is in root mode, which is not the normal operation mode. Drones, on the other hand, are normally equipped with an embedded computer that can run an operating system such as Linux. Wi-Fi transceivers are easily configured in ad hoc mode within the Linux operating system.

4.4 Our Proposed Approach: Evolutionary Deployment and Hill Climbing-Based Movements

The primary objective of the proposed approach is to provide communication to the maximum number of victims. The main assumption is that certain information about the disaster scenario, such as likely positions of victims, might exist. We define a target waypoint as a region where we prioritize coverage based on evidence of possible victims. Normally, a waypoint is defined by its GPS coordinates. This information can be obtained using satellite images or phone calls made by people who saw the disaster. With this information, we determine a first deployment of drones in the disaster area. In this case, drones work in coverage mode and are placed in the positions that guarantee the maximum possible coverage to the waypoints. That is, initially the drones form a connected ad hoc mesh network, such that all drones are reachable from every other drone. We note, however, that this initial deployment can be far from optimal since we only have partial information on the disaster scenario. We define a knowledge level K, which is the percentage of known waypoints or victims. (More details on K exist in the simulation results in Sect. 4.5.) Once the initial deployment is carried out, drones can obtain more accurate real-time information and then use this info to adjust the initial deployment. Drones will then adapt their positions according to the victims' positions. In this phase of the proposed approach, where the drones work in exploration mode, the objective is to cover unexplored areas to find new victims. If a new victim is discovered, the positions of drones are updated to expand coverage to the newly found victims, while still maintaining a connected ad hoc mesh network.

Figure 4.2 shows the anticipated timeline of our proposed approach. First, the disaster event occurs, and the time for collecting information begins. Then, the drones are sent to the disaster area. Once the drones arrive at the disaster area,

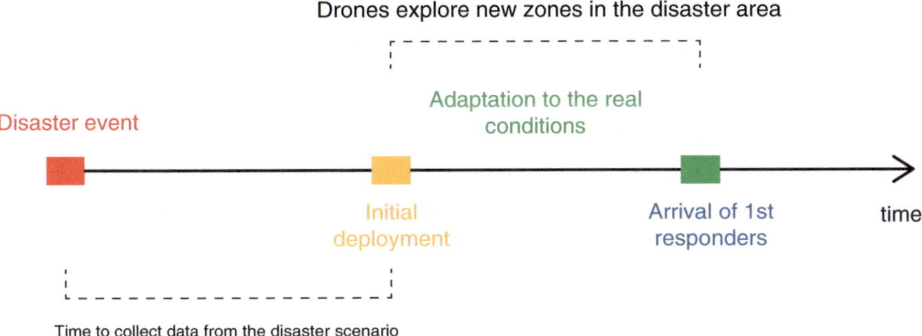

Fig. 4.2 Timeline of the proposed approach

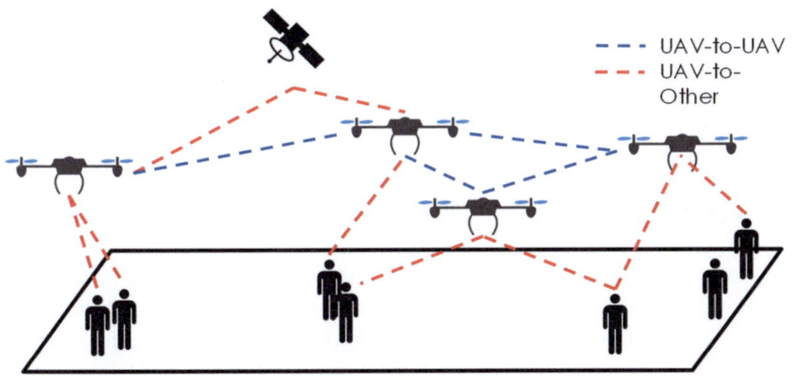

Fig. 4.3 An Example of drones' deployment and communications

the initial deployment is carried out using the collected information (see Fig. 4.2). After the initial deployment, the drones start to work in a distribute way to explore new zones (adaptation to the real/changing conditions in Fig. 4.2). Finally, the first responders arrive at the disaster area.

Figure 4.3 illustrates an example network deployment and communications that are possible with a set of drones in a disaster scenario. Figure 4.3 also shows how the drones form a mesh network (blue dotted lines) and how the drones provide communication services to the victims.

Another possible way to collect the locations of victims in a disaster area is to use their cell phones to sense the number of neighbors (other victims) that are in their vicinity. This would be especially useful if the batteries of some victims were off. We note that we do not consider the battery lifetime of flying drones in our approach. Obviously, battery lifetime is an important issue that will limit the lifetime of the mesh network formed by the drones. One simple solution to a drone with a low battery level is to replace the drone with a new drone that has a high battery level.

Furthermore, it is important to highlight that our proposed deployment works both in a centralized and distributed way. On the one hand, the global search algorithm used in the initial deployment has to be executed by a central unit with high computing resources. In general, evolutionary algorithms demand high computing resources since they need to complete many operations to deeply explore a significant part of the search space. On the other hand, the local search algorithm used in the adaptation to the changing conditions uses only local information, i.e., nodes exchange information that is then used by the HCA. For this reason, the HCA demands much less computing power than the genetic algorithm (GA).

4.4.1 Initial Deployment

The initial deployment problem basically consists of finding the optimal positions of a number of drones working in coverage mode so that they can provide communication services to the target waypoints and surrounding areas.

4.4.1.1 Formal Definition of the Problem

Given an array of target positions P, such that each element of the array is a tuple (x, y) that represents the Cartesian coordinates of a victim, and $|P| = z$, where z represents the number of victims and/or waypoints that need to be covered, the objective consists of finding the most optimal positions (x, y)[1] of N drones equipped with wireless transceivers that form a connected ad hoc mesh network to cover the maximum number of positions in P.

A network or graph is connected iff there is a path between any pair of nodes. In our case, the nodes are drones. A drone i is connected to another drone or a victim j iff $d_{ij} < r$, where d_{ij} is the Euclidean distance between the two nodes and r is the drone's radio transmission range.[2] Drones can calculate the value of d_{ij} using their GPS coordinates. A further restriction is the number of victims, v, that can be covered by a drone. We consider $v \leq V_m$, where V_m is the maximum number of clients that can be served simultaneously (which is defined by the wireless chipset used).

The above-defined problem is like the well-known set cover problem, which has been demonstrated to be an NP-hard problem [34]. In other words, there is no algorithm that solves the problem in polynomial time. We need an algorithm that provides us an optimal or quasi-optimal solution in a reasonable time. Thus, we

---

[1] We consider that drones are placed in 2D space.

[2] The drone's transmission area is assumed to be a perfect circle of radius r according to the unit disk model.

assume the availability of high computing resources for the initial deployment since the initial deployment would be planned in the headquarters of the first responders.

4.4.1.2 An Evolutionary Algorithm Approach

We propose the use of evolutionary algorithms (EAs), i.e., a genetic algorithm, to solve the initial deployment problem, which consists of finding the optimal positions of drones to cover the maximum possible number of victims. Our GA uses the global information collected on the disaster scenario before the drones arrive.

EAs use nature-inspired evolution strategies to solve complex problems. They are based on the Darwinian theory of evolution [35], which describes the capacity of biological systems to modify their genetic material to adapt to a changing environment and ensure their survival. EAs are iterative heuristics that evolve a set of candidate solutions, represented as individuals that are grouped in a population. The study and design of new EAs are a very active research topic in the artificial intelligence research community [36, 37].

The basic idea of a GA is that an initial population composed of potential solutions (individuals) evolves over time by generating better solutions based on the previous generation of solutions. In general, a GA composed of g generations and each generation contains n individuals. Here, g and n are design parameters. After the gth generation, the resulting population is composed of the n best individuals (solutions) found in the execution of the GA. In our initial deployment, the solutions are the positions of the drones in the disaster scenario P. The GA begins with an initial population, which is selected randomly. Each individual of the initial population represents a different potential solution. In the target optimization problem, a potential solution is given by the drones' coordinates in the disaster area. This is called the "chromosome structure" in evolutionary computation and defines the nature of an individual potential solution. For example, for n drones, we obtain a potential solution representation such as the one shown in Fig. 4.4.

Where Xi and Yi in Fig. 4.4 represent the x and y coordinates of drone i. The size of the population, that is the number of individuals or potential solutions, normally depends on the number of variables that form an individual solution [39]. It is important to highlight that the potential solutions in the initial optimization problem should meet the requirement of forming a connected network; otherwise, the solution is not valid. Figure 4.5 illustrates two possible potential solutions. The upper solution is a valid one because the drones form a connected network. Conversely, the lower solution does not comply with our connectivity requirement since the red drone is isolated from the others. As a result, it is considered an invalid

X1, Y1	X2, Y2	X3, Y3	...	Xn, Yn

Fig. 4.4 Genetic information of an individual (or potential) solution

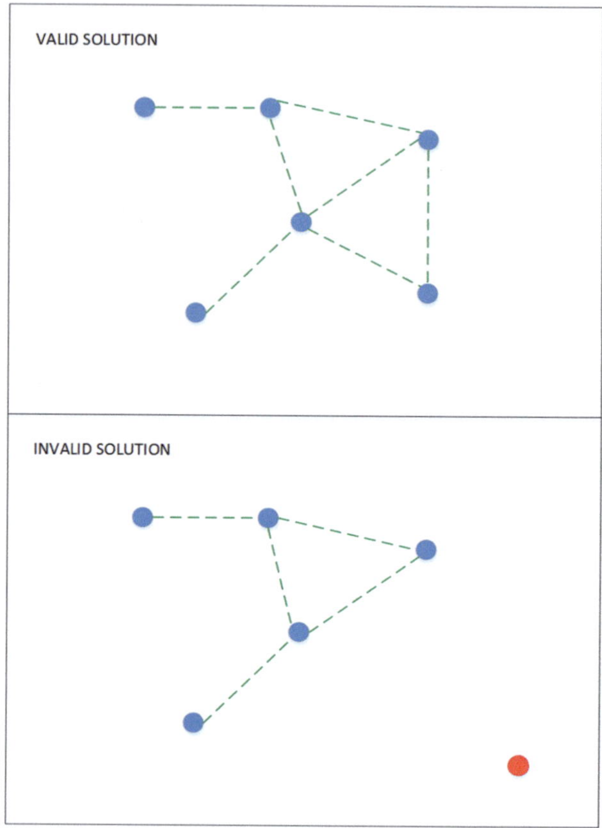

Fig. 4.5 Representation of potential solutions

solution. For every possible solution, the connectivity requirement has to be checked before evaluating its fitness function (i.e., the quality of the solution).

Once we have generated the initial population, each individual solution is evaluated using a fitness function that determines the quality of the given solution. In the initial deployment problem, the quality of each individual solution is determined as follows:

$$f = k_1\, V_c + k_2\, C, \tag{4.1}$$

where V_c is the number of victims covered by the drones and C is the total number of possible connections among the drones and victims. That is, if a victim is in the overlapping area of two drones' transmission areas, it will be counted twice by C. The objective of C is to measure the redundancy of communications among victims and drones. Redundant connections are advantageous in the event that a drone fails, since another drone can provide service to the victims that were originally covered by the malfunctioning drone. Consequently, if two possible solutions cover the same number of victims, the term C will determine the best solution. The terms k_1 and

k_2 are two constants that can be adjusted to vary the importance of each term. The value of C can be much higher than V_c because C does not consider the limitation imposed on the number of clients that a drone can handle. We, therefore, want to give higher importance to the term V_c; thus, we propose that k_1 and k_2 meet the following condition:

$$k_2 = \frac{k_1}{|I| * |I| * |P|},\tag{4.2}$$

where $|I|$ is the number of drones and $|P|$ is the number of victims or waypoints to be covered. Notice that $|I| * |I| * |P|$ is the total number of possible connections among victims and drones. Using Eqs. (4.1) and (4.2) we obtain a rich range of solutions. Thus, we calculate other metrics related to the robustness of the solutions. For example, the number of extra connections E can be calculated as $E = C - V_c$. This metric represents the robustness of the solution in terms of possible communication failures. If a communication between a drone and a victim fails, the victim can still be covered by another drone. Moreover, we can also define the robustness percentage of the solutions as $R\,(\%) = \frac{E}{V_c} 100$.

Before evaluating a solution with Eq. (4.1), we should first guarantee that it is valid. In other words, we must first verify that the drones form a connected network. If an individual solution is not a valid solution, it is penalized with a fitness rating of -1 (see Fig. 4.5 for an example of an invalid solution). Consequently, the fitness of a solution will be determined as follows:

$$f = k_1\,V_c + k_2\,C \quad \text{if valid}$$
$$f = -1 \quad \text{otherwise}\tag{4.3}$$

The drones can sense other drones that are within their radio transmission ranges to maintain neighbor tables. That is, they can know the topology of the network at any instant in time. Notice that this is easily achieved by the exchange of Hello packets. The connectivity algorithm starts with an empty list of reachable nodes. Then, the algorithm in the list adds each reachable node from a given node. At the end of the algorithm, the length of the list must be equal to the number of drones to verify all drones are reachable. If all the nodes are reachable, the algorithm returns $Net = 0$; otherwise, the algorithm returns $Net = -1$.

We now illustrate the two main components of the above fitness function with the two examples in Figs. 4.6 and 4.7. For both examples, there are 6 drones and 7 victims, and we set $k_1 = 100$. We then determine k_2 heuristically as

$$k_2 = \frac{100}{6 * 6 * 7} = 0.40$$

Figure 4.6 shows two possible solutions that cover different numbers of victims. According to the aforementioned k_1 and k_2 values, we calculate the quality of both solutions f_1 and f_2 using Eq. (4.3) as

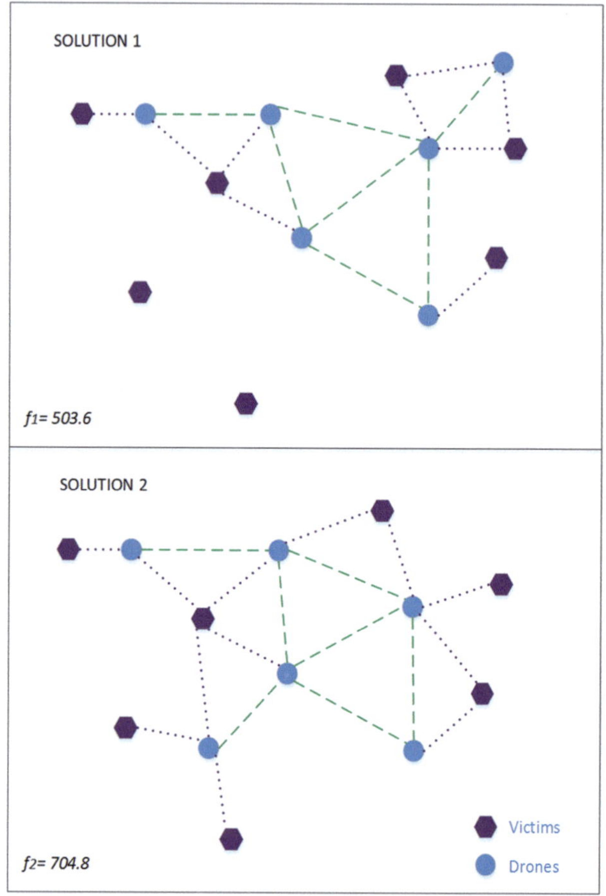

Fig. 4.6 Illustration of the importance of k_1 in the fitness function, i.e., Eq. (4.3)

$$f_1 = 5 * 100 + 9 * 0.4 = 503.6$$

$$f_2 = 7 * 100 + 12 * 0.4 = 704.8$$

We observe that the term k_1 makes a large difference in the two solutions. Notice that even in the hypothetical case, where f_2 has more connections, the term k_1 gives much more importance to the number of victims.

Figure 4.7 represents two more possible solutions for the same scenario, such that the two solutions give coverage to the same number of victims. That is, in both solutions V_c is 7. The total number of connections, C, is 9 connections in Fig. 4.7 for solution 1 and 12 connections for solution 2. Using the calculated value of k_2, we determine the quality of these two solutions (f_3 and f_4) using Eq. (4.3) as

$$f_3 = 7 * 100 + 9 * 0.4 = 703.6$$

$$f_4 = 7 * 100 + 12 * 0.4 = 704.8$$

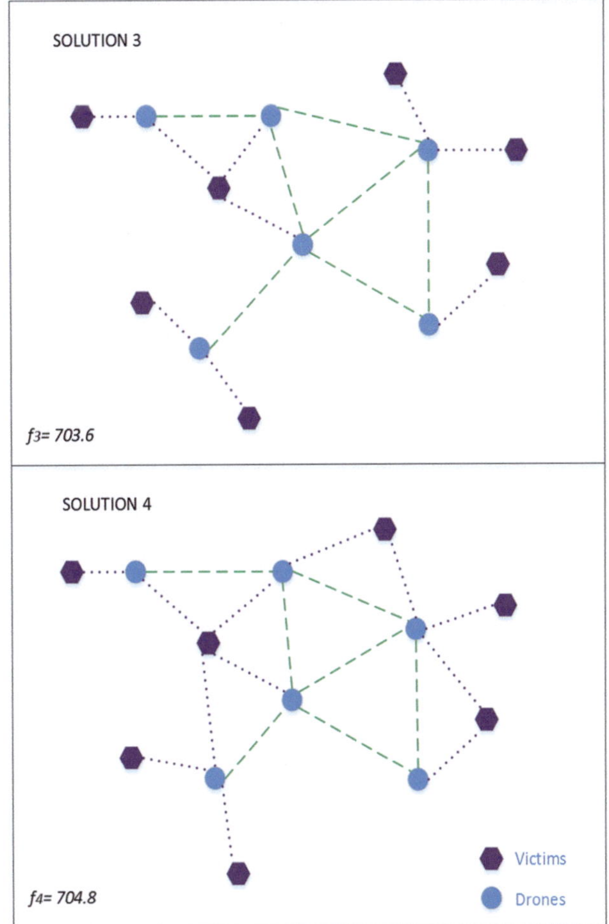

Fig. 4.7 Illustration of the importance of k_2 in the fitness function, i.e., Eq. (4.3)

As shown in the obtained values of f_3 and f_4, solution 4 is slightly better than solution 3 because it is more robust against possible communication failures in the communications between drones and victims. If a victim is covered by two drones, he or she can retain communication abilities even if one of the drones fails. Consequently, the term k_2 only becomes important when two possible solutions cover the same number of victims. Otherwise, the term k_1 dominates the quality of the solution.

We now highlight the connectivity requirement with another example (Fig. 4.8). Again, two possible solutions are considered, but, in this case, solution 5 does not comply with our connectivity requirement. If we apply Eq. (4.3) with the previous values of k_1 and k_2, we see that solution 5 would have a fitness score of 904.4 if it were a valid solution; however, it is not valid; hence it receives a score of -1. Solution 6 receives a comparatively high score of 704.8.

Fig. 4.8 Connectivity requirement

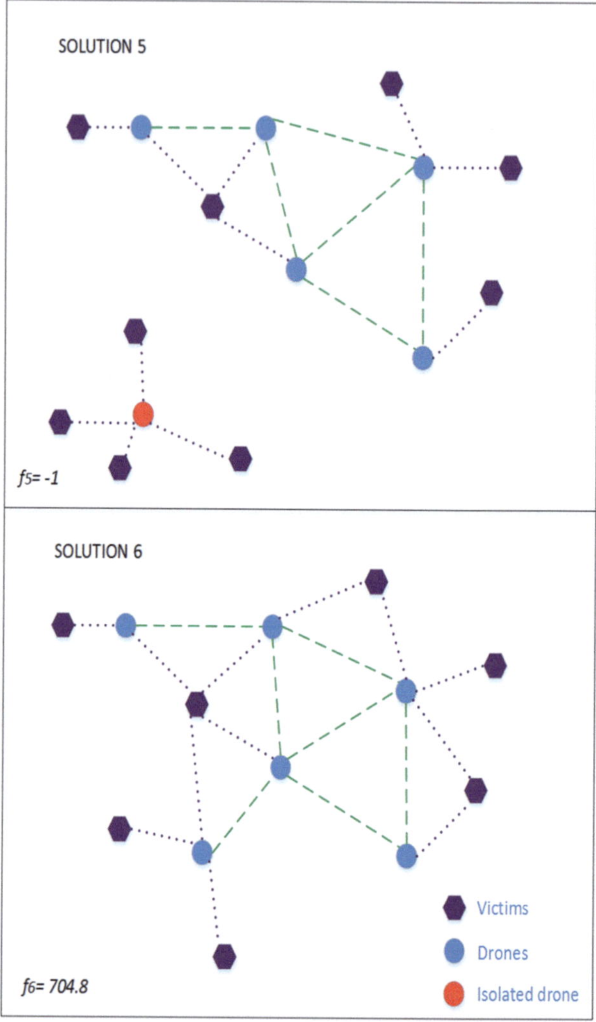

We now describe how the GA evolves. The evolution of the population is determined by two genetic operations included in the GA. There are several methods to select a new parent including tournament, roulette, and elitism [41]. A parent is a previously obtained solution that is used to generate new solutions by applying genetic operations over it. The main idea behind the selection mechanism is to select parents based on their quality. As a rule, the higher the quality of an individual, the higher the probability of it being selected as a parent. Consequently, all the individuals of a population are sorted according to their quality and assigned corresponding probabilities for being one of the parents of the next generation. It is worth recalling that every individual of the population is a potential solution.

Table 4.1 Parent selection mechanism based on probability roulette

Individual/potential solution	Quality	Probability of being selected
S1	700.5	700.5/1651.5 = 0.43
S2	500.7	500.7/1651.5 = 0.30
S3	300.2	300.2/1651.5 = 0.18
S4	150.1	150.1/1651.5 = 0.09

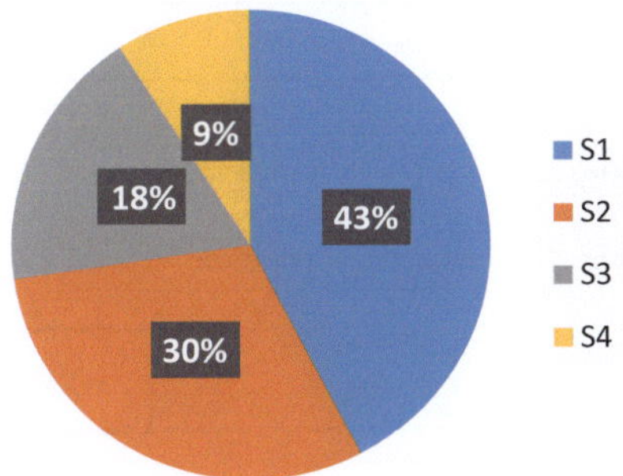

Fig. 4.9 Probability roulette

We illustrate parent selection using probability roulette, which is the method used in our simulations. Table 4.1 contains the population composed of four individuals or possible solutions (S1, S2, S3, S4) to a hypothetical initial deployment optimization problem. The quality column in Table 4.1 shows the quality of each solution, and the rightmost column contains the probability of each individual being selected as a parent. These probabilities are based on the qualities. Figure 4.9 is a graphic illustration of the probabilities in Table 4.1.

We also use an elitism mechanism to guarantee that the best individuals of each generation pass directly to the next generation (or "offspring" in evolutionary computation). Using the previous example and assuming 25% elitism (this value is just an example), then one of the individuals included in Table 4.1 will pass directly to the new generation. According to the qualities included in Table 4.1, the solution S1 will go directly to the next generation. In short, the best positions for the drones found so far always pass to the next generation in the evolution.

Once we have selected the parents, the main genetic operators to create new individuals (including crossover and mutation) are applied on the selected parents. Via the crossover operation, the genetic information of the two parent solutions is mixed. There are many different crossover operators, such as one-point crossover, two-point crossover, and uniform [41]. Figure 4.10 illustrates a single crossover operation between two selected parents that represent two potential solutions composed of five drones. The crossover point is selected randomly.

The mutation operator involves modifying the genetic information of an individual solution. Again, different mutation operators can be applied, such as Gaussian and shuffle indexes [41]. Figure 4.11 illustrates the mutation operation for a selected parent. Only part of the individual is modified to generate a new one.

Crossover and mutation operators are applied according to certain probabilities p_c and p_b. The main objective of crossover is to combine the genetic information of

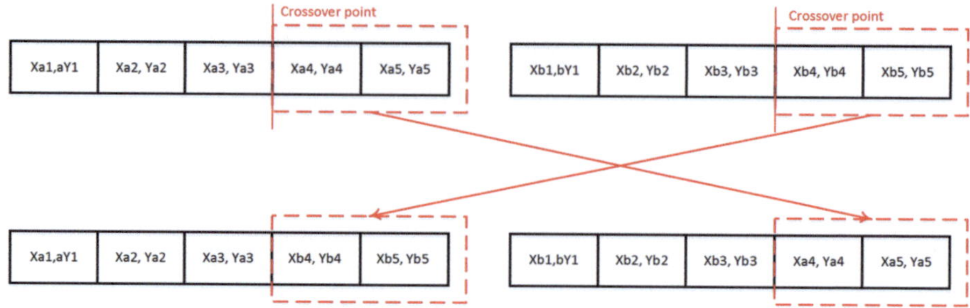

Fig. 4.10 Single crossover operation

Fig. 4.11 Mutation operation

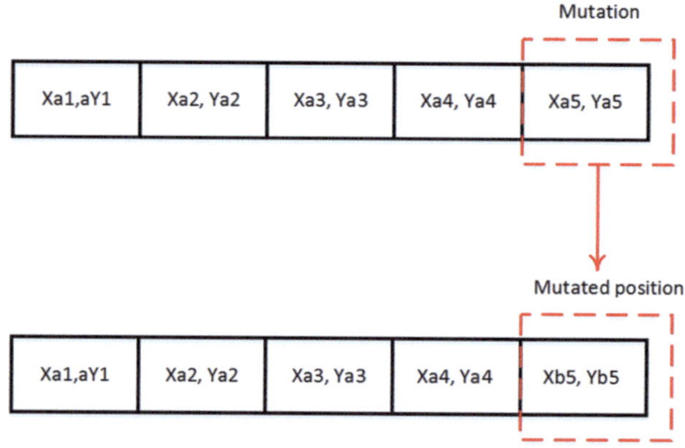

two individuals to determine whether the individual solution improves the existing solutions. The main goal of mutation is to explore new locations in the research space and introduce genetic diversity to avoid getting stuck in local optima. Once we have determined the composition of the new population, it is evaluated until the stopping criterion is reached. In general, two possible stopping criteria are to fix the number of evolution generations and to stop evolution once the average population fitness stagnates. In the former case, the population evolves for a given of number of generations without considering the quality of the individuals. In the latter case, if the average individuals' quality does not change for a given number of consecutive generations, we assume that the most optimal value has been reached. Whenever the algorithm stops, the resulting population contains the best solutions found for the evolution of the previous generations.

4.4.2 Adaptation to the Changing Conditions

Once the drones have moved to their initial deployment positions according to the coordinates obtained from our evolutionary algorithm, they switch to exploration mode to try to improve the current configuration. That is, the drones move in new directions without abandoning coverage of known victims or disconnecting the network. A new configuration is considered better if it has a higher fitness than the previous configuration as calculated by Eq. (4.3). In general, there are two ways to improve the initial configuration: increasing V_c by finding new victims or increasing the number of connections C.

4.4.2.1 Formal Definition of the Problem

Given an array of drone positions P' determined by the initial deployment optimization problem, we seek to adapt the positions of the drones to the changing conditions of the disaster scenario by exploring new positions in the surrounding area. The goal is to improve the global quality of the solutions achieved by the genetic optimization. Again, the drones must always form a connected network, and the number of victims that a given drone can handle is given by $V_{m\cdot}$.

The present problem is also an NP problem since it is very similar to the initial deployment problem. Consequently, we need a heuristic algorithm that provides us a quasi-optimal solution in reasonable time. That is, we want a new solution determined in real time because the drones will already be providing communication services to the victims. In addition, we cannot use global information; the drones must locally determine whether the new solution is possible. Although this problem is somewhat like the previous optimization problem, we should not use a GA for several reasons. First, GAs require high computing power to evaluate many potential solutions over the course of the population's evolution. In addition, such massive evaluations without high computing power would require prohibitive computing time, which is not acceptable in real-time scenarios such as the one described here. Second, the operators used by GAs, such as crossover and mutation, require global information, which cannot be possible in a distributed network such as the mesh network formed by the drones.

4.4.2.2 A Local Search Algorithms Approach

We propose to use a local search algorithm, i.e., the HCA [42], to allow the drones to adapt their configuration to observed conditions. The HCA is a mathematical optimization technique that falls into the category of local search optimization algorithms. It is an iterative algorithm that begins with an arbitrary solution to a problem and then attempts to find a better solution by incrementally changing a single element of the solution. If the incremental change produces a better solution, then the change is made to the new solution; this process is then repeated until no further improvements are found. In our optimization problem, a better solution means that the new solution increases the fitness function, Eq. (4.3). Although the HCA is a simple algorithm, it is able to solve complex NP problems [16]. The HCA begins with a random position or potential solution, which is considered at this stage

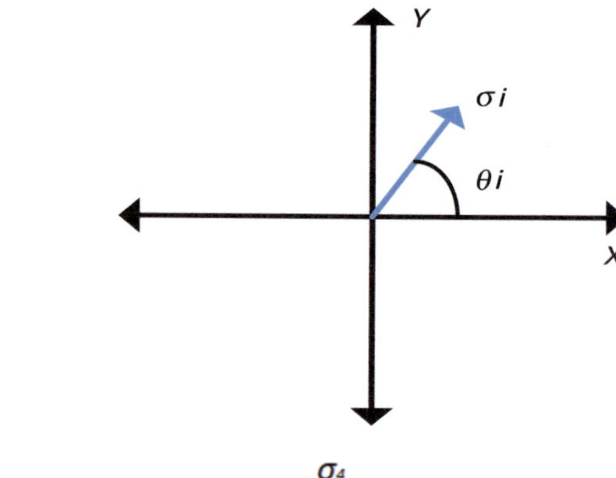

Fig. 4.12 A drone's moving angle (ϑ_i) and speed (σ_i) in the HCA

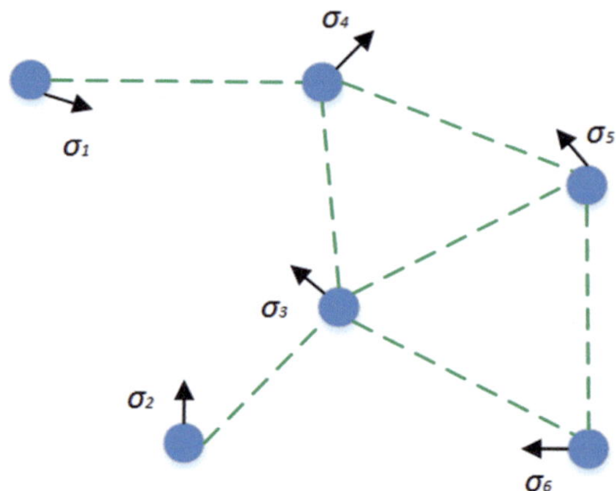

Fig. 4.13 Example of moving directions in the HCA

as the best solution. The structure of a potential solution is the list of drone positions for a disaster. In the adaption to the changing conditions optimization problem, the initial position is the best position obtained by the previous initial optimization problem P'.

Then, the algorithm selects a direction in which the position of drones is moved. The moving direction is given by two parameters, i.e., the vector of angles ϑ and the vector of speeds σ for the set of drones. The vector ϑ_i contains the angle used by a given drone i to determine the moving direction at speed σ_i. Figure 4.12 illustrates how the moving angle is measured. The vector of speeds contains the magnitudes of the speed vectors.

When a new moving direction has to be selected because the current direction does not improve the current situation, then every drone randomly selects a new moving angle within the interval $[0, 2\pi]$ and a speed magnitude within the interval $[0, Vmax]$. Figure 4.13 illustrates the moving directions for six drones. Once the moving directions are selected, the drones move to their new positions, and the quality of the new solution is evaluated.

If the new solution improves the current best position, then the best position is updated. The algorithm will also maintain the same direction of movement as long as the solution keeps improving. Otherwise, the drones go back to the previous positions, and the algorithm selects a new direction. The algorithm continues iterating until the maximum number of interactions is reached. It is important to recall that the new positions of drones must comply with the connectivity requirement; otherwise, the solution will be penalized with a negative fitness.

4.5 Simulation and Results

This section presents our simulation results for our proposed approach, which we evaluate in a disaster scenario under different conditions. Conducting real experiments to evaluate the performance of the mesh network formed by drones in real disaster scenarios is very difficult for several reasons. First, it implies a high investment in terms of hardware. Second, flying drones requires specific licenses and insurances, which depend on local laws and regulations. Third, emulating the changing conditions of victims in a disaster is complicated, and many people have to be involved. Therefore, most research today regarding the performance of drones with wireless capabilities is based on simulation. Consequently, it is important to emulate the conditions of disaster scenarios as realistically as possible [40]. The algorithms have been coded in Python 2.7, and the Distributed Evolutionary Algorithms in Python (DEAP) module (version 1.0.2) was used to implement the GA [43]. The code is available at [4].

4.5.1 Disaster Scenario Description

The simulation of the disaster scenario attempts to emulate a rural disaster in a countryside area; thus, the drones move freely throughout the whole disaster scenario. In this case, the victims are static and are distributed around the four corners (four big clusters) of a square disaster area. The value of K, which represents the percent of known victims in a real scenario, will depend on the data collected in the aftermath of the disaster event. There are several mechanisms to obtain the location of victims including satellite images and phone calls from people who live around the disaster area. Figure 4.14 represents the scenario considered versus the value of K. Notice that the resulting scenarios for $K = 0.4$ and $K = 0.6$ are some of the most difficult because many unknown victims are in the same corner of the disaster area; also, because of the lack of knowledge within this corner area, the initial deployment will not place drones here. Consequently, drones will likely be located at the opposite corners, and they must explore the uncovered areas during the exploration mode phase while maintaining the connectivity of the mesh network.

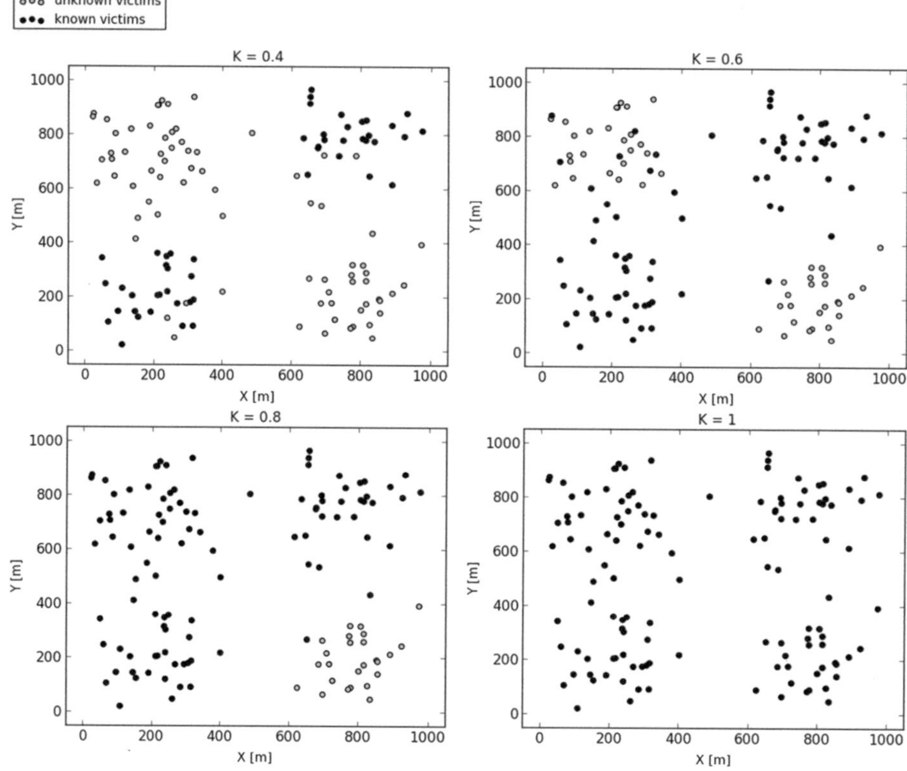

Fig. 4.14 Disaster scenario considered vs. level of knowledge (K)

Table 4.2 Number of known and unknown victims vs. level of knowledge (K)

K value	0	0.2	0.4	0.6	0.8	1
Nº known victims/waypoints	0	25	50	75	100	125
Nº unknown victims/waypoints	125	100	75	50	25	0

Table 4.2 contains the number of known and unknown victims with respect to the value of K. We designate a victim[3] as unknown when we do not know the victim's location in the disaster scenario. The initial deployment problem consists of finding the most optimal positions of the drones to provide services to the known victims (black points in Fig. 4.14). Once the drones are deployed according to the initial optimization problem, we begin exploration mode to adapt to the changing conditions and improve the current solution. Since the drones are in the disaster area, the locations of all victims can be found by the drones. When drones move

---

[3]We refer to a victim as a waypoint, as the drone is likely to find more victims at the waypoint location.

Table 4.3 Global simulation parameters

Simulation parameter	Value
Total N° victims	125
Mobility of victims	Static
Total N° drones	10
V_m	15
Drone's transmission radio range	250 m
Disaster area	1000 m × 1000 m

according to the local search algorithm, they find the positions of new victims (unknown victims in the initial deployment problem, highlighted as yellow points in Fig. 4.14).

4.5.2   Simulation Setup

The global simulation parameters used in our simulations are included in Table 4.3.

We use a radio transmission range of 250 m because it is the standard value used in most ad hoc network studies that are based of IEEE 802.11 a/b protocols [38]. We use the disk connectivity model such that a drone can communicate with another drone or a victim if they are within radio transmission range of each other. We consider the propagation of the signals to be ideal. This assumption fits well since the scenario considered is a rural scenario with few obstacles. The connection threshold, denoted by V_m, is 15 since we assume that the drones are equipped with commercial Wi-Fi dongles. We note that a different value of V_m only changes the number of needed drones to cover the victims, i.e., the complexity of the problem would be similar. We assume that victims are static during the simulations. We do not expect huge differences with mobile victims since our drones adapt to the changing conditions in real time. We run 50 different trials of the GA. The ideal number of trials depends on two factors: (1) computing power available and (2) the time available before the drones arrive at the disaster location.

The configuration parameters used by the GA implementation are included in Table 4.4. We use a population size of 100 individuals (potential solutions), which guarantees enough exploration during the first stage of the algorithm [39]. We note a population of 100 individuals and 50 different trials means we consider up to 5000 random deployments. The selection mechanism used is the probability roulette mechanism, where the probability of being selected as a parent is proportional to the individual's quality (see Sect. 4.4 for more details). We use Eq. (4.3) to determine the quality of the individual solutions, and 90% of the offspring are generated by applying genetic operators (crossover and mutation) to parents from the previous generation. The remaining 10% of the offspring are members of the previous generation that pass directly to the next generation because of the elitism mechanism. We note that the percentage of solutions we should pass directly to a new generation should not be high to guarantee enough exploration. The elitism

Table 4.4 Configuration parameters for the GA

Configuration parameter	Value
Population size	100
% Generated by crossover and mutation	90%
% Generated by elitism	10%
Type of selection	Roulette
Type of crossover	Two points
Crossover probability	80%
Type of mutation	Shuffle indexes
Mutation probability	20%
Fitness function	Eq. (4.3)
$N^{\underline{o}}$ Generations	100

Table 4.5 Configuration parameters for the HCA

Configuration parameter	Value
Simulation time	5000 s
Simulation step time	1 s
Drone's maximum speed	10 m/s
Fitness function	Eq. (4.3)

mechanism, however, is suitable for ensuring that the best solutions are not lost during the execution of the GA due to poor genetic combinations among the selected parents. In the proposed approach, some solutions obtain negative fitness due to our connectivity requirement. Regarding the mutation and the crossover probabilities, we should select a high value for the crossover probability to guarantee high exploration of the search space and, on the other hand, a low value for the mutation probability to slightly modify some solutions. For example, a 0% of crossover probability means that we do not generate new possible solutions based on the existing ones. Furthermore, a high value of mutation probability means that every solution is highly modified and, thus, not desirable because we shift solutions from good areas to worse areas in the search space. We run 100 generations of the GA. We checked to ensure this number is suitable for the convergence of the optimization problem.

Table 4.5 contains the configuration parameters used by the HCA implementation. The simulation time refers to the time during which drones are trying to adapt their positions to the current victims' locations. The selected simulation time guarantees convergence, which means that the best solution has been found. The positions of the drones are updated with 1 s time steps, and the maximum drone speed is 10 m/s.

4.5.3 Results and Analysis

We evaluate the proposed approach under different levels of knowledge. We fix the number of drones as ten and vary the value of K within the interval [0, 1] in steps

Table 4.6 Simulation results for the initial deployment (GA) with 10 drones vs. K value

Metric	$K = 0.2$	$K = 0.4$	$K = 0.6$	$K = 0.8$	$K = 1$
Maximum fitness ($R\%$)	2502.0 (900%)	5001.600 (300%)	7401.696 (186.49%)	9902.024 (155.55%)	11702.008 (101.7%)
Minimum fitness	2501.912	4501.232	6801.816	8601.864	10001.632
Mean	2501.996	4819.412	7121.6264	9416.022	10851.786
Standard deviation	0.016	139.57	134.163	254.576	393.109

Table 4.7 Simulation results for the adaptation of the changing conditions (HCA) with 10 drones vs. K value

Metric	$K = 0.2$	$K = 0.4$	$K = 0.6$	$K = 0.8$	$K = 1$
Best fitness (R%)	3002.256 (840%)	6802.056 (277.94%)	8902.048 (186.51%)	10002.128 (166%)	11902.048 (115.13%)

of 0.2. This number of drones guarantees that all victims can be covered since the connection threshold V_m is *15* (see Table 4.3) and the number of victims is 125. This analysis evaluates our proposed approach under different values of K. Tables 4.6 and 4.7 contain the simulation results obtained by our proposed approach under different values of K. The results in Tables 4.6 and 4.7 represent the best individuals (sets of positions) from the evolution of the GA and HCA, respectively. In Table 4.6, the maximum fitness values represent the quality of the best solution achieved by the initial deployment problem using Eq. (4.3) and considering 50 different trials of the GA. Similarly, the minimum fitness values represent the minimum value obtained by the initial deployment using Eq. (4.3) and the 50 trials. Table 4.6 also includes both the mean and standard deviation calculated from the best solutions of the 50 different trials. Let us use an example to understand what the fitness values represent in Tables 4.6 and 4.7. A fitness value of 11702.008 means that the number of victims found (V_c) is 117 since $k_1 = 100$ in Eq. (4.3). With respect to the second term in the fitness function (Eq. 4.3), $k_2 = 0.0085$ for 10 drones. Consequently, the 2.008 in the fitness of this solution is related to the number of possible communications between the drones and the victims[4] (see Eq. (4.1) in Sect. 4.4.1.2 for more details). We calculate C using Eq. (4.1) as $C = \frac{f - k_1 V c}{k_2}$. In this case, the value of C is 236; we then calculate the number of extra communications (E) between drones and victims as $236 - 117$ (number of victims covered) $= 119$. This means that the drones may still cover all victims even if 119 avenues of communication fail. We also calculate the robustness percentage of the solutions as *R(%)* (see Sect. 4.4.1.2 for more details). In our example, *R(%) = 101.7*. In the following tables, we include the *R(%)* for the maximum fitness value obtained for the sake of comparison. Not surprisingly, as the number of victims increases, the value of R decreases since more victims have to be covered by the same number of drones.

---

[4]This example is valid for the rest of the tables in this chapter that show simulation results.

Table 4.8 Simulation results for the initial deployment (HCA) with 10 drones vs. K value

Metric	$K = 0.2$	$K = 0.4$	$K = 0.6$	$K = 0.8$	$K = 1$
Maximum fitness	2501.368 (584%)	4700.984 (161.70%)	7101.472 (159.15%)	8802.440 (246.59%)	10601.960 (131.13%)
Minimum fitness	0.0	1300.216	2700.528	2800.416	5401.784
Mean	1368.374	2992.713	4613.287	5901.563	7777.876
Standard deviation	971.001	755.560	1049.606	1507.164	1199.7618

Table 4.7 contains the best solution achieved by the HCA during the 5000 s considered. Notice that we do not include the results for $K = 0$ in Tables 4.6 and 4.7 since that would refer to the number of known victims being zero; in this case, it is more suitable to apply the local search optimization directly. Another important feature of our proposed approach is that the adaptation to the changing conditions, which uses the HCA, finds more robust solutions when K is large ($K > 0.6$). Notice that for the highest value of K, 95.2% of the victims are found and 24% for the lowest value of K.

In order to justify the application of a GA for the initial deployment problem, Table 4.8 shows the results of using the HCA for the initial deployment problem. If we compare the results in Tables 4.6 and 4.8, we clearly see that the GA outperforms the HCA for the initial deployment. Specifically, for all the tested K values, the results of the GA are better than the ones obtained by the HCA. The differences are remarkable in terms of standard deviation. This indicates that, with the application of the GA, we achieve both better and less dispersed results.

Table 4.9 summarizes our simulation results obtained by both the genetic and the HCA. It also compares the results obtained by the two optimization algorithms. The second leftmost column contains the fitness of the best solution for the GA in the initial deployment problem, the number of victims covered, and the percentage of known victims that are covered. It is worth recalling that, for the initial deployment problem, the GA only uses the positions of known victims for the deployment. The next column includes the fitness of the best solution achieved by the HCA in the adaptation to the real conditions problem, the number of victims covered, and the percentage of the total victims that are covered (known and unknown victims). The rightmost column includes the absolute value of the difference between the fitness values of the two algorithms, the difference in number of victims, and the percentage with respect to the total number of victims. The difference between both algorithms significantly depends on the value of K. For high values of K, the GA determines the optimal positions of drones using complete (or almost complete) knowledge so there is little difference between the two algorithms, i.e., only 1–2 new victims are found for K values higher than 0.6. On the other hand, for low values of K, the HCA explores new areas in the disaster scenario away from known victims, improving the results achieved by the initial deployment (e.g., 18 new victims for $K = 0.4$).

Figure 4.15 shows the drones' final positions for the disaster scenario considered when $K = 0.6$. The dotted red lines represent the communication links among the deployed drones. The dotted yellow lines represent the possible communication

Table 4.9 Comparison of the results obtained by the genetic algorithm and the HCA for 10 drones and different values of K

K value	Genetic algorithm (victims) (% with respect to Nº of known victims)	Hill climbing (victims) (% with respect to total Nº of victims)	Fitness difference (victims) (% with respect to total Nº of victims)
0.2	2502.0 (=25 victims) (100%)	3002.256 (=30 victims) (24%)	500.256 (=5 victims) (4%)
0.4	5001.600 (=50 victims) (100%)	6802.056 (=68 victims) (54%)	1800.456 (=18 victims) (14.4%)
0.6	7401.696 (=74 victims) (99%)	8902.048 (=89 victims) (71%)	1500.386 (=15 victims) (12%)
0.8	9902.024 (=99 victims) (99%)	10002.128 (=100 victims) (80%)	100.104 (=1 victim) (0.8%)
1	11702.008 (=117 victims) (94%)	11902.048 (=119 victims) (95%)	200.040 (=2 victims) (1.6%)

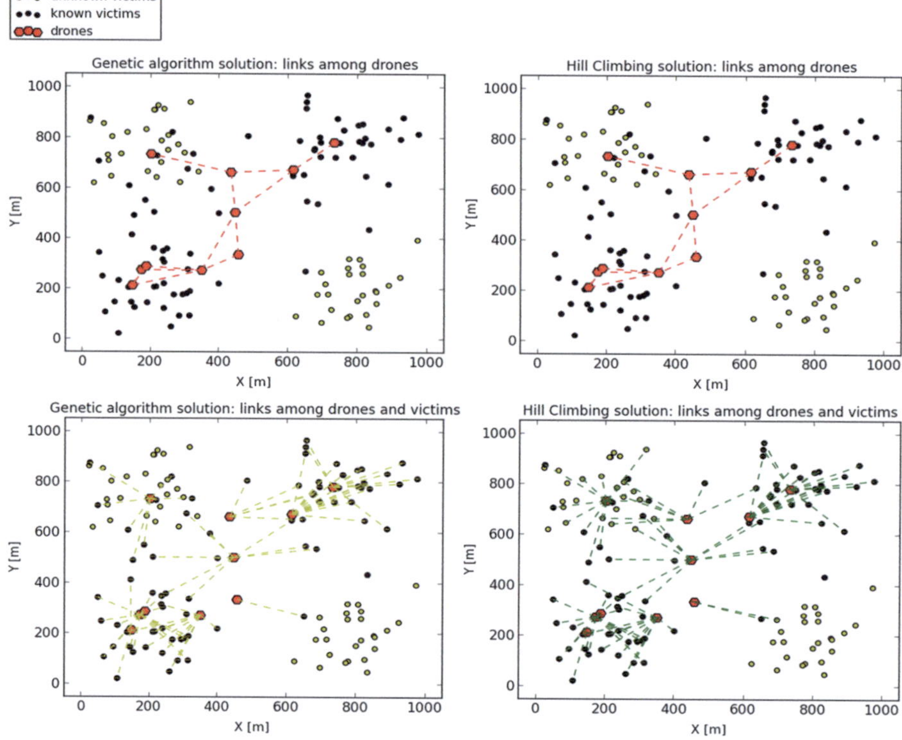

Fig. 4.15 Drones' positions and communication links for 10 drones and $K = 0.6$

links among the drones and the victims, after the initial deployment of drones based on the GA. Finally, the dotted green lines represent the possible communication links among the victims and the drones resulting from the adaptation to the changing conditions with the HCA. We observe that several victims in the lower right corner are not covered. Unfortunately, in this corner, the number of unknown victims is much higher than the number of known victims (only one victim is known). Consequently, during the initial deployment, none of the drones were placed at this corner, and the HCA cannot send a drone to this corner without breaking the connectivity requirement.

4.6 Conclusions

In this chapter, we propose the use of drones as 0th responders in a multi-UAV network to provide communication services to victims in disaster scenarios. The goal of the 0th responders is to arrive at the disaster scenario before the first responders. We divide their deployment and movements into two phases. The first phase uses a GA to find the most optimal positions of the drones based on certain information previously collected from the disaster scenario. The second phase consists of adapting the positions to the conditions of the disaster scenario and exploring new areas to find more victims. The second phase relies on the use of a local search algorithm, such as the HCA. We have evaluated the proposed approach in simulated rural disaster scenarios involving 125 static victims under different conditions in terms of levels of knowledge. The obtained simulation results are very satisfactory in most of cases, covering an important percentage of victims in the disaster scenario considered. In case that full knowledge of the scenario is assumed and 10 drones are used, then 95.2% of the victims are covered by the proposed approach. Suitable results are also obtained for other vales of knowledge. It has been demonstrated that search power of the GA can be improved by incorporating an exploitative algorithm like the HCA. As future work, we plan to evaluate the proposed approach with other meta-heuristics such as particle swarm optimization and ant colony optimization.

References

1. Ochoa, S. F., Neyem, A., Pino, J. A., & Borges, M. (2007). Supporting group decision making and coordination in urban disasters relief efforts. *Journal of Decision Systems, 16,* 143–172.
2. Asimakopoulou, E., & Bessis, N. (2010). *Advanced ICTs for Disaster management and threat detection: Collaborative and distributed frameworks.* IGI Publishing, ISBN: 978-1615209873.
3. Muench, M. (2011). *Classification of emergency scenarios.* TU Darmstadt Pro Seminar.
4. Reina, D. G., Toral, S. L., Barrero, F., Bessis, N., & Asimakopoulou, E. (2013). The role of ad hoc networks in the internet of things. In N. Bessis, F. Xhafa, D. Varvarigou, R. Hill, & M. Li (Eds.), *Internet of things and inter-cooperative computational technologies for collective intelligence* (Vol. 460). Berlin: Springer.
5. Günes, M., Reina, D. G., García-Campos, J. M., & Toral, S. L. (2017). *Mobile Ad Hoc network protocols based on dissimilarity metrics.* Cham: Springer. https://doi.org/10.1007/978-3-319-62740-3_5.
6. Reina, D. G., Askalani, M., Toral, S. L., Barrero, F., Asimakopoulou, E., & Bessis, N. (2015). A survey on multihop ad hoc Networks for disaster response scenarios. *International Journal of Distributed Sensor Networks.* Article ID 647037.
7. Ketterling, H. A. (2004). *Introduction to digital professional mobile radio.* Boston: Artech House.
8. Zheng, Y., Chen, S., & Ling, H. (2015). Evolutionary optimization for disaster relief operations: A survey. *Applied Soft Computing, 27,* 553–566.
9. Merwaday, A., & Güvenç, I. (2015). *UAV assisted heterogeneous networks for public safety communications.* IEEE Wireless Communications and Networking Conference Workshops (WCNCW), pp. 329, 334.
10. Richards, M. D., Whitley, D., & Beveridge, J. R. (2005). *Evolving cooperative strategies for UAV teams.* In Proceedings of the Genetic and Evolutionary Computation Conference, pp. 1721–1728.
11. Reina, D. G., Tawfik, H., & Toral, S. L. Multi-subpopulation evolutionary algorithms for coverage deployment of UAV-networks. *Ad Hoc Networks.* https://doi.org/10.1016/j.adhoc.2017.09.005.

12. Sharma, V., Reina, D. G., & Kumar, R. (2017). HMADSO: A Novel Hill Myna and desert sparrow optimization algorithm for cooperative rendezvous and task allocation in FANETs. *Soft Computing*. https://doi.org/10.1007/s00500-017-2686-4.

13. Sánchez-García, J., García-Campos, J., Toral, S., Reina, D. G., & Barrero, F. (2016). An intelligent strategy for tactical movements of UAVs in disaster scenarios. *International Journal of Distributed Sensor Networks, 12*(3), 1–20.

14. Agustín-Blas, L. E., Salcedo-Sanz, S., Vidales, P., Urueta, G., & Portilla-Figueras, J. A. (2011). Near optimal citywide WiFi network deployment using a hybrid grouping genetic algorithm. *Expert Systems with Applications, 38*, 9543–9556.

15. Oda, T., Sakamoto, S., Spaho, E., Barolli, A., Barolli, L., & Xhafa, F.. (2014). *Node placement in WMNs using WMN-GA system considering uniform and normal distribution of mesh clients.* Eighth International Conference on Complex, Intelligent and Software Intensive Systems, pp. 120–127.

16. Chang, X., Oda, T., Spaho, E., Ikeda, M., Barolli, L., & Xhafa, F. (2014). *Node placement in WMNs ssing WMN-HC system and different movement methods.* Eighth International Conference on Complex, Intelligent and Software Intensive Systems, pp. 1148–1153.

17. Konak, A., Buchert, G. E., & Juro, J. (2013). A flocking-based approach to maintain connectivity in mobile wireless ad hoc networks. *Applied Soft Computing, 13*, 1284–1291.

18. Abdelkhalek, O., Krichen, S., & Guitouni, A. (2015). A genetic algorithm based decision support system for the multi-objective node placement problem in next wireless generation network. *Applied Soft Computing, 33*, 278–291.

19. Reina, D. G., Toral, S. L., Bessis, N., Barrero, F., & Asimakopoulou, E. (2013). An evolutionary computation approach for optimizing connectivity in disaster response scenarios. *Applied Soft Computing, 13*, 833–845.

20. Chakraborty, N., Sycara, K.. (2010). *Reconfiguration algorithms for mobile robotic networks.* IEEE International Conference on Robotics and Automation (ICRA), pp. 5484–5489.

21. Tekdas, O., Plonski, P. A., Karnad, N., Isler, V. (2010). *Maintaining connectivity in environments with obstacles.* IEEE International Conference on Robotics and Automation (ICRA), pp. 1952–1957.

22. Aschenbruck, N., Gerhards-Padilla, E., & Martini, P. (2009). Modeling mobility in disaster area scenarios. *Performance Evaluation, 66*, 773–790.

23. N. Aschenbruck, Frank, M., Martini, P., Tölle, J. (2004). *Human mobility in MANET disaster area simulation – A realistic approach.* In 29th Annual IEEE International Conference on Local Computer Network (LCN'04).

24. Huang, Y., He, W., Nahrstedt, K., & Lee, W. C. (2008). CORPS: Event-driven incident mobility model for first responders. IEEE Military Communications Conference (MILCOM 08), pp. 1–7.

25. Conceição, L., & Curado, M. (2013). *Modelling mobility based on human behaviour in disaster areas.* In Proceedings of the 11th International Conference of Wired/Wireless Internet Communication (WWIC), pp. 56–59.

26. Torres, R., Mengual, L., Marban, O., Eibe, S., Menasalvas, E., & Maza, B. (2012). A management ad hoc networks model for rescue and emergency scenarios. *Expert System with Applications, 39*, 9554–9563.

27. Uddin, M. Y. S., Nicol, D. M., Abdelzaher, T. F., & Kravets, R. H.. (2009). *A post disaster mobility model for delay tolerant networking.* In Proceedings of the Winter Simulation Conference, pp. 2785–2796.

28. Pascarella, D., Venticinque, S., Aversa, R., Mattei, M., & Blasi, L. (2015). Parallel and distributed computing for UAVs trajectory planning. *Journal Ambient Intelligence and Humanized Computing, 6*, 773–782.

29. Agrawal, V., Lightner, C., Lightner-Laws, C., & Wagner, N. A bi-criteria evolutionary algorithm for a constrained multi-depot vehicle routing problem. *Soft Computing*. https://doi.org/10.1007/s00500-016-2112-3.

30. Arzamendia, M., Gregor, D., Reina, D. G., & Toral, S. L. (2017). An evolutionary approach to constrained path planning of an autonomous surface vehicle for maximizing the covered area of Ypacarai Lake. *Soft Computing*. https://doi.org/10.1007/s00500-017-2895-x.
31. Camp, T., Williams, B. (2002). Comparison of broadcasting techniques for mobile ad hoc networks. In Proceeding of the ACM International Symposium on Mobile Ad Hoc Networking and Computing, pp. 194–205.
32. Reina, D. G., Johnson, P., Toral, S. L., & Barrero, F. (2015). A survey on probabilistic broadcast schemes for wireless ad hoc networks. *Ad Hoc Networks, 25*, 263–292.
33. Boukerche, A., Turgut, B., Aydin, N., Ahmad, M. Z., Boloni, L., & Turgut, D. (2011). Routing protocols in ad hoc networks: A survey. *Computer Networks, 55*, 3032–3080.
34. Uriel, F. (1998). A threshold of ln n for approximating set cover. *Journal of the ACM, 45*, 634–652.
35. Darwin, C. (1859). *The origin of species by means of natural selection: Or, the preservation of favored races in the struggle for life.* London: John Murray.
36. Klüver, J., & Klüver, C. (2016). The regulatory algorithm (RGA): a two-dimensional extension of evolutionary algorithms. *Soft Computing, 20*, 2067–2075.
37. Díaz-Manríquez, A., Toscano, G., & Coello-Coello, C. A. Comparison of metamodeling techniques in evolutionary algorithms. *Soft Computing.*https://doi.org/10.1007/s00500-016-2140-z.
38. IEEE Standard for Information Technology – Telecommunications and Information Exchange Between Systems – Local and Metropolitan Area Networks – Specific Requirements – Part 11: Wireless LAN Medium Access Control (MAC) and Physical Layer (PHY) Specifications. IEEE Std 802.11-2007 (Revision of IEEE Std 802.11-1999).
39. Alander, J. T. (1995). *On optimal population size of genetic algorithm.* Proceedings CompEuro 1992. Computer systems and software engineering, 6th Annual European Computer Conference, pp. 65–70.
40. Ficco, M., Avolio, G., Palmieri, F., & Castiglione, A. (2016). An HLA-based framework for simulation of large-scale critical systems. *Concurrency and Computation: Practice and Experience, 28*, 400–419.
41. http://deap.gel.ulaval.ca/doc/default/api/tools.html#operators. Last access July 2015.
42. Gent, I. P., & Walsh, T. (1993). *Towards an understanding of hill-climbing procedures for SAT.* Proceedings of the eleventh national conference on Artificial intelligence (AAAI;93), pp. 28–33.
43. Fortin, F., De Rainville, F., Gardner, M., Parizeau, M., & Gagne, C. (2012). DEAP: Evolutionary algorithms made easy. *Journal of Machine Learning Research, 13*, 2171–2175.
44. https://github.com/Dany503/DeploNet

Chapter 5
Detection of Obstructive Sleep Apnea Using Deep Neural Network

Mashail Alsalamah, Saad Amin, and Vasile Palade

5.1 Introduction

Sleep apnea is a potentially common sleep disorder in which a person's breathing may have one or more pauses during sleep. These pauses may continue from a few seconds to several minutes and may occur hundreds of times during the night. If the obstruction to breathing is total and continues for 10 or more seconds, then this case is called *apnea*. During the sleep apnea, the brain and the rest of the body may not get enough oxygen. As a result, the quality of sleep is poor, which makes the patient tired during the day [1]. In addition, it is considered a risk factor for morbidity and mortality due to its long-term effect on the cardiovascular system [2].

Sleep apnea typically is classified into three types: obstructive sleep apnea (OSA), central sleep apnea (CSA), and mixed sleep apnea (MIX). OSA is the more common form of apnea; it is caused by a blockage of the airway and is generally associated with a reduction in blood oxygen saturation, whereas in CSA the airway is not blocked, but the brain fails to signal the muscles to breathe, due to instability in the respiratory control center, while MIX occurs due to transition between long periods of OSA and brief intervals of CSA [3].

Traditionally, sleep-related breathing disorders are diagnosed by visual observation of polysomnography (PSG) signals. PSG is a sleep test that is performed at special laboratories. It consists of recording various signals including the breath airflow, respiratory movement, oxygen saturation, body position, electroencephalogram (EEG), electrooculogram (EOG), electromyogram (EMG), and electrocardiogram (ECG) [4].

Even though PSG become the standard diagnostic tool for sleep disorder cases, there are some problems related to its implementation which make it expensive

M. Alsalamah (✉) · S. Amin · V. Palade
Faculty of Engineering and Computing, Coventry University, Coventry, UK

© Springer International Publishing AG, part of Springer Nature 2018
M. M. Alani et al. (eds.), *Applications of Big Data Analytics*,
https://doi.org/10.1007/978-3-319-76472-6_5

and time consuming. For example, its execution requires the patient to sleep in a sleep laboratory for one or two nights, in the presence of technicians. Furthermore, patients must maintain a particular position throughout the night with special equipment attached to their bodies during measurement. Hence these limitations put a barrier to PSG acceptance and reduced its diagnostic power. Therefore, the need for a simpler alternative detection method has been arising. Automated methods that use the artificial intelligence algorithms can solve PSG problems, since it is easier and faster to detect OSA cases. Furthermore, due to the increasing interest of a wearable and portable sleep quality monitoring system for home care which requires the use of minimum number of channels, OSA detection based on single-lead ECG is gaining keen interest in the sleep research community. In this light, there have been a number of algorithms proposed to tackle the problem of automatic OSA detection using ECG patterns obtained during PSG studies using machine learning techniques.

ECG recording is one of the simpler and efficient technologies in sleep disorders detection. In 2000, Computers in Cardiology (CINC) and PhysioNet organized a competition to highlight the potential use of the ECG signals in diagnosing sleep apnea. They hosted a challenge where various research teams introduced a number of different methodologies for sleep apnea detection using the ECG.

PhysioNet provided free access to the database of ECG recordings and an automatic web-based scoring program. The recordings were arranged in three classes, as follows:

(1) Class A (apnea): recordings in this class contain at least 1 h with an apnea index of ten or more and at least 100 min with apnea during the recording. The learning and test sets each contain 20 class A recordings.

(2) Class B (borderline): recordings in class B contain at least 1 h with an apnea index of five or more and between 5 and 99 min with apnea during the recording. The learning and test sets each contain five class B recordings.

(3) Class C (control): recordings in class C contain fewer than 5 min with apnea during the recording. The learning and test sets each contain ten class C recordings.

The competition consisted of two challenges. The first challenge was to identify the recordings in the test set with sleep apnea (class A) and the normal recordings (class C). Assignments for class B were not scored. The score was the total number of correct classifications of class A and class C, so that the maximum possible score was 30. The second challenge was to label each minute in all 35 test recordings as either containing apnea (A) or not (N). In this challenge, all 35 test recordings were scored [5].

The aim of this study is to propose a novel scheme for OSA detection based on features of ECG signals. This scheme is a hybrid algorithm that combines the deep neural network (DNN) with the decision tree. The classification process in this proposed scheme consists of two phases; the first phase uses DNN for minute-based classification, and then the output of this phase is fed into a decision tree model in order to perform the second phase, class identification. In addition to the proposed scheme, a comparative study of the most used classification methods, which have not been used with the same features and dataset, adopted in the literature is done.

The rest of this chapter is organized as follows. In Sect. 5.2, we summarize the related work in the literature. Section 5.3 contains an overview of the system and details the paper methodology. In Sect. 5.4, we present the experimentations that we have done and the obtained results. Section 5.5 concludes our study and lists possible extensions to this work.

5.2   Related Work

Since PhysioNet/CinC challenge, many methods using the ECG signal to diagnose OSA have been proposed. The algorithms in this research area were divided into two types, some for apnea classes' identification and others for the minute-by-minute apnea classification.

Regarding the first challenge, several algorithms using different methods were developed to identify apnea class. For example, in [6, 7, 8] authors made use of spectral analysis of heart rate variability (HRV) to identify apnea class and achieved 30 correct score out of 30 (without class B consideration). While authors in [9, 10] used the Hilbert transform to extract frequency information from the heart rate signal and achieved a score of (28/30). Authors in [6, 8, 11] achieved the top three ranks in the PhysioNet's challenge on the subject of the minute-by-minute quantification. They reached an accuracy of 89.4%, 92.6%, and 92.3%. In addition to HRV, authors made use of different features derived from ECG signals like ECG pulse energy [8], R wave amplitude using power spectral density (PSD) [6], and T wave amplitude using the discrete harmonic wavelet transform [11].

Since the challenge, various methods have been proposed in the automated OSA detection on the same PhysioNet Apnea_ECG dataset. Khandoker et al. [12] employed wavelet-based features and K-nearest neighbor (KNN) classifier to achieve an accuracy of 83%. Xie et al. [13] extracted features from ECG and saturation of peripheral oxygen (SpO2) signals and employed classifier combination such as AdaBoost with decision stump and bagging with REPTree where the classification accuracy was 77.74%.

Many studies show that detection of obstructive sleep apnea can be performed through HRV and the ECG signal. Quiceno-Manrique et al. [14] proposed a simple diagnostic tool for OSA with a high accuracy (up to 92.67%) using time-frequency distributions and dynamic features in ECG signal. Moreover, based on spectral components of heart rate variability, frequency analysis was performed in [10] using Fourier and wavelet transformation with appropriate application of the Hilbert Transform, where the sensitivity was 90.8%. In addition, in [15] a bivariate autoregressive model was used to evaluate beat-by-beat power spectral density of HRV and R peak area, where the classification results showed accuracy higher than 85%. The technique in this work also relies on features of the ECG signal. In 2012, Laiali Almazaydeh et al. proposed an automated classification algorithm based on support vector machine (SVM) using statistical features extracted from ECG signals for both normal and apnea patients based on heart rate variability (HRV), with an accuracy of 96.5% [16].

Even though the aforementioned studies achieved relative satisfactory performance on apnea detection and quantification, some important aspects have to be highlighted. First, the proposed approaches either identify apnea class or detect the presence or absence of each minute of ECG data. To the best of our knowledge, only authors in [17, 18] addressed both apnea detection and quantification for each patient recording, but both identify only two classes, not three (class B is excluded). Second, various features are extracted from the RR intervals without careful investigation, causing predictors in the selected classifier to be more redundant. At the same time, feature extraction and selection from such high-dimensional feature spaces would require a large amount of computational resource, which is not attainable for most wearable devices and is also inconvenient for their wider application in home-based diagnosis since the modern healthcare system is required to assist physicians to quickly determine subjects' status with which physicians can provide a quick pre-diagnosis for the subjects. Hence, to address these issues, this study proposes a novel OSA screening approach to achieve a satisfactory performance using fewer features under limited capacities of wearable devices.

5.2.1 Deep Neural Networks

Deep learning is currently one of the most important active research areas in machine learning. It has attracted extreme attention from researchers due to its potential in a wide range of active applications such as object recognition [19, 20], speech recognition [21], natural language processing, theoretical science, medical science [22], etc.

Deep learning has also been used commercially by companies like Google, Facebook, Twitter, Instagram, Apple, and others to provide better services to their customers. For example, Apple's Siri, the virtual personal assistant in iPhones, provides a large variety of services including weather reports, sport news, answers to user's questions, and other different services by exploiting the high capability of deep learning approaches. Also, other famous companies make use of deep learning techniques to improve their services, like Microsoft's real-time language translation in Bing voice search [23], IBM's brain-like computer, and Google in Google's translator, Android's voice recognition, Google's street view, and image search engine.

Deep learning can be defined as a machine learning technique that makes use of neural networks which are linked in a hierarchical architecture that in turn uses several layers to produce an output. Each layer receives input from a layer below, transforms its representations, and then propagates it to the layer above.

Inspired by the biological nature of human brain mechanisms for processing of natural signals, deep neural networks are representation learning methods with multiple levels of representation.

The expression "deep" is used because the depth of the network is greater when compared to the more conventional neural networks, which are sometimes called shallow networks. In most conventional learning methods, a simple network with one hidden layer may achieve acceptable performance for performing a specific task, but by applying a deep architecture with more hidden layers, higher efficiency can be achieved. This is because each hidden layer extracts more features from the previous layer and creates its own abstract representation. Therefore, to resolve more complicated features, we have to add more hidden layers, which make deep learning capable of learning latent information.

This concept of deep networks is similar to hierarchical neural networks like the neocognitron model proposed by [24] but with some differences in the architecture and learning algorithm. The neocognitron model uses winner-take-all unsupervised learning, whereas learning in deep networks is done by the back-propagation algorithm.

The term "deep" learning was coined as a contrast to the "shallow" learning algorithms which have fixed and usually single layer architecture. The "deep" learning architectures are compositions of many layers of adaptive nonlinear components. It is expected that by analogy with the mammal brain capable of storing information on several layers of different abstractions, these multilayer architectures will bring improvement to future learning algorithms. However, simple training of the neural networks with up to two or three multiple hidden layers has shown an improvement, but further increases in the number of layers did not provide any significant improvement and in some case the results were worse. The existing algorithms have faced the problem of the local minimum, and it has been reported that the generalization of such gradient-based methods has become worse with a larger number of layers. Several papers have also shown that supervised training of each separate layer does not give a significant improvement in results compared to regular multilayer learning. Later development has gone in the direction of the intermediate feature representation for each new layer. Deep learning networks and training algorithms using this approach have achieved significant results in the multiple real-life applications [25], including computer vision, audio signal processing, natural language processing, and so on. In some fields of study, they are still considered to be among the best available approaches.

The successful examples of the deep neural networks for supervised learning mainly exploit two different approaches and their possible combinations which include both the special structure of the network in terms of neuron connections with hierarchically organized feature transformations applied to their results (i.e., convolutional neural networks) and also multilayer networks with feature representations for each layer learned with unsupervised learning technique followed by parameter tuning of the network using a regular supervised learning technique.

Convolutional neural networks were specially designed for visual object recognition, and they were based on the modern ideas concerning the working of human visual perception. Owing to their special structure, they are easier to train than conventional fully connected networks. The convolutional network designed by LeCun is currently the best known for character recognition. This network has two types of layers which include convolutional and subsampling layers. Neurons are associated with the fixed two-dimensional positions on the input image with the weights assigned to the rectangular patch of the image. This locality principle allows the learning of local features such as edges and shapes. These features are later hierarchically combined by the higher layers of the network. The nodes corresponding to the learned features are later copied with the same weights to different network positions. This is done based on the assumption that the same low-level features can be met in the different locations of the image. It also helps to overcome difficulties caused by small input distortions and invariant transformations like shift or rotation. In general weights sharing decrease learning complexity of the networks requiring optimize less parameters but with more effective representation. The neurons and weights produced by this method constitute a feature map which is applied to the different parts of the image. The convolutional layer is composed of several feature maps which allows the computing of several types of features at each location. It is followed by the subsampling (pooling) layer which performs averaging and subsampling of the computed features to reduce the overall complexity by the further application of nonlinear (sigmoid) transformation. The procedure is repeated hierarchically to the next layer with rectangular patches of the neurons on previous layers being assigned to the neurons on the higher layer. Several groups of these layers are used to constitute the overall network. Theories as to why the performance of the convolutional networks is better than that of the multilayer neural networks have not yet been fully developed. One hypothesis was the fact that each neuron uses only a small part of the input space helps to avoid gradient diffusion common for fully connected multilayered networks.

Deep learning networks first appeared in 2006 with deep belief networks and were followed by stacked autoencoders and other types of algorithms all based on the same principle. Each layer of the network is pre-trained with unsupervised learning algorithms which learn nonlinear transformations from the output of the previous layer (intermediate feature space). For example, the autoencoder is a simple neural network applied to the unlabeled input with a minimum of restrictions on the number of activated neurons on the hidden layer. In other words it tries to approximate the input data but with fewer parameters. As a result such an encoder can discover any interesting patterns in the data. Previously this technique had been used only to remove noise from the data. Deep belief networks are based on the restricted Boltzmann machines which in fact are acyclic graphs which attempt to discover any probability distributions or dependencies in the data. Unsupervised training is intended to explain the variations in the input data. After the processing of each layer is completed, networks initialized in this manner are fine-tuned with the regular gradient-based learning method. The experiments on the different datasets have shown that unsupervised pre-training provides some sort of regularization factor which minimizes variance and introduces a bias toward the configuration of the input feature space, which is useful for unsupervised learning. This is different from random parameters initialization where the probability of selecting parameters which lead toward the local minimum of the optimization criteria is very high, in which case unsupervised learning provides the algorithm with an insight which leads to better generalization.

5.3   Methodology

This work is based on the ECG signal features to detect sleep apnea. Figure 5.1 represents the block diagram for the methodology proposed in this study.

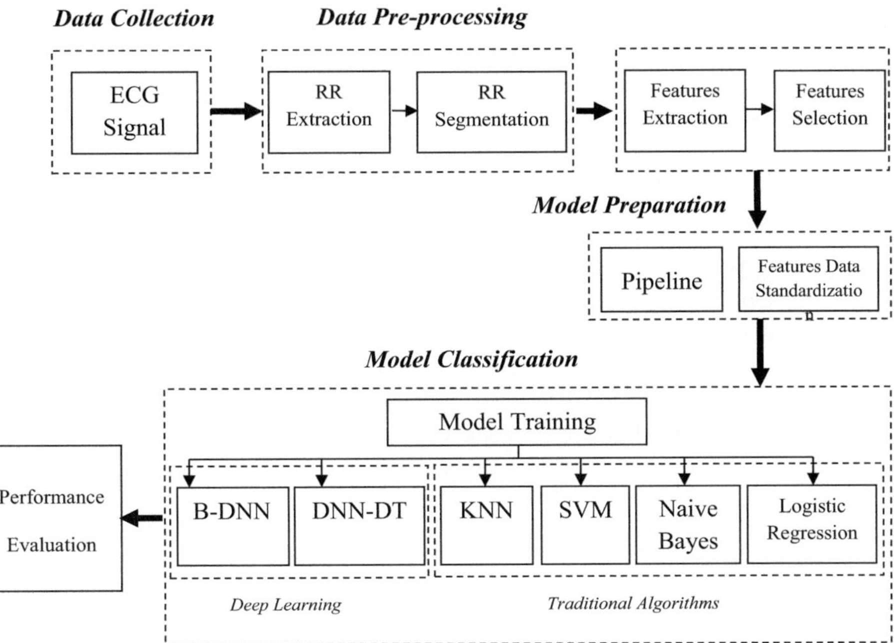

Fig. 5.1 Block diagram of the proposed methodology

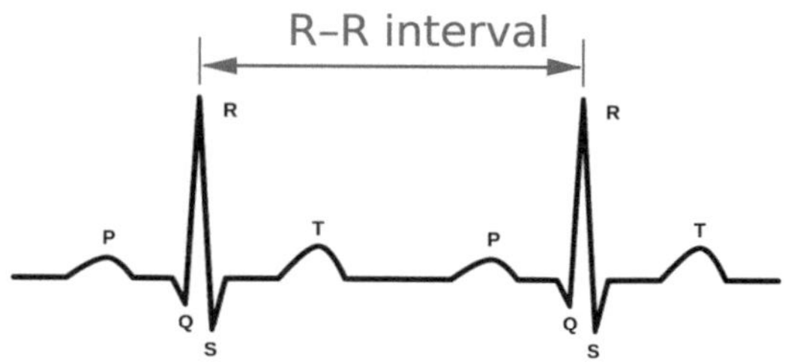

Fig. 5.2 Calculation of RR intervals

5.3.1 Data Collection

The experimental data used in this study was obtained from the PhysioNet Apnea_ECG dataset [26]. The Apnea_ECG dataset contains ECG recordings for 70 different patients with OSA (classes a, b, c). Recordings vary in duration from slightly less than 7 h to nearly 10 h each. However, only 35 of these recordings contain minute-wise apnea annotations, which indicate the presence or absence of apnea during each minute of ECG data. ECG signals are sampled at 100 Hz with 12-bit resolution.

5.3.2 Data Preprocessing

5.3.2.1 RR Intervals Detection

The features used in our experimentation were all metrics based around RR intervals. An RR interval is defined as the time between two consecutive R peaks (Fig. 5.2), which in turn are defined as the maximum amplitude of a given QRS complex. QRS is defined as the deflections in an electrocardiogram (EKG) tracing that represent the ventricular activity of the heart. It is the combination of the Q wave, R wave, and S wave and represents ventricular depolarization. The normal duration of the QRS complex is 0.08 and 0.10 s. When the duration is between 0.10 and 0.12 s, it is intermediate or slightly prolonged. A QRS duration of greater than 0.12 s is considered abnormal [27]. These metrics were chosen because RR intervals have been shown to be a telling indicator of HRV, which is a known byproduct of sleep apnea [28].

WaveForm DataBase (WFDB) Software Package [29] was used to extract specific signals and annotations from the ECG recordings files. WFDB is a large collection of software to access PhysioBank and for viewing, annotation, and analysis of signals. Also it includes command-line tools for signal processing and automated analysis. Specifically, we used WFDB package to extract RR intervals using "ann2rr" command.

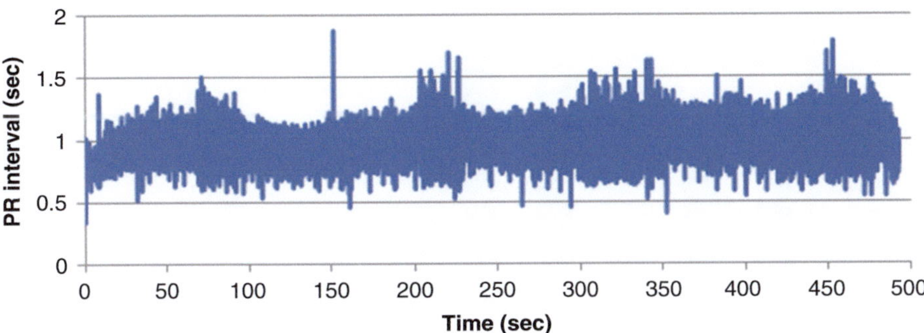

Fig. 5.3 Extracted RR intervals from (a01) file

5.3.2.2 RR Intervals Segmentation

As each ECG recording in the PhysioNet database was annotated per minute, the extracted RR intervals are segmented on a minute-by-minute basis according to the annotations. Therefore, RR intervals were calculated for each minute at each file, which implies that we have about 17,003 RR records (35 file * file length (450–550 min)). Figure 5.3 presents the extracted RR values for one of the data set files (a01 file).

5.3.3 *Feature Extraction*

For each segment of the RR intervals obtained from the previous preprocessing phases, statistical features could be extracted and fed into the classification model for the possible classification of apnea events. Each feature vector was computed based on 60 s of ECG data, as each minute-wise annotation indicates the presence or absence of apnea at the beginning of the following minute. The following ECG features, which are the most common features used in the literature [4, 30] for apnea detection, are calculated:

1. Mean of the RR interval
2. Median of RR intervals
3. Standard deviation (SD) of the RR interval
4. The NN50 measure (variant 1), defined as the number of pairs of adjacent RR intervals where the first RR interval exceeds the second RR interval by more than 50 ms
5. The NN50 measure (variant 2), defined as the number of pairs of adjacent RR intervals where the second RR interval exceeds the first RR interval by more than 50 ms

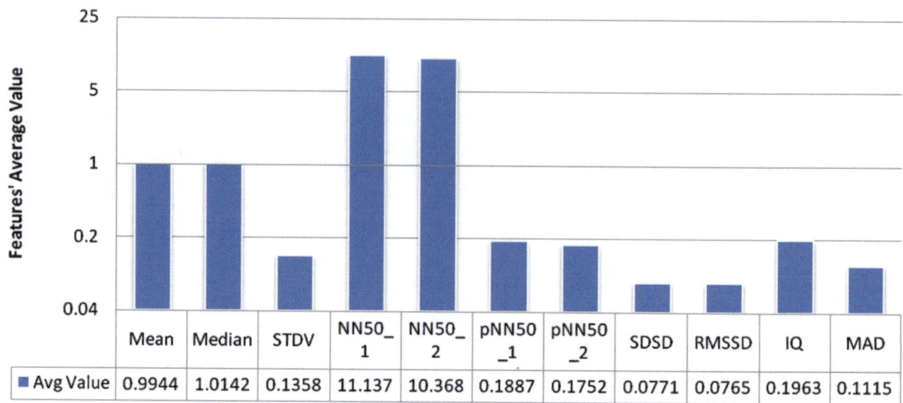

Fig. 5.4 Extracted features from RR intervals of (a01) file

6. The PNN50_1 measures, defined as NN50 (variant 1) measure divided by the total number of RR intervals
7. The PNN50_2 measures, defined as NN50 (variant 2) measure divided by the total number of RR intervals
8. The SDSD measures, defined as the SD of the differences between adjacent RR intervals
9. The RMSSD measures, defined as the square root of the mean of the sum of the squares of differences between adjacent RR intervals
10. Interquartile range, defined as difference between 75th and 25th percentiles of the RR interval value distribution
11. Mean absolute deviation values, defined as mean of absolute values by the subtraction of the mean RR interval values from all the RR interval values in an epoch

Figure 5.4 presents the average value of the extracted features from RR intervals that are extracted from a01 file of the dataset.

5.3.4 Feature Selection

There exists an important trade-off in performing feature extraction: the use of more features will lead to higher levels of classification accuracy but comes at the price of taking longer to perform apnea detection in real time. Ranking is crucial for achieving better-quality analysis results. Thus, we sought to identify the optimal features subsets from the original features set to minimize the size of our features vector while still being able to classify sleep apnea with high accuracy.

In this phase, the features, which have the strongest effect on prediction, are selected. This stage scores the attributes according to their correlation with the classified apnea class. It selects the most informative attributes. In total, 11 features

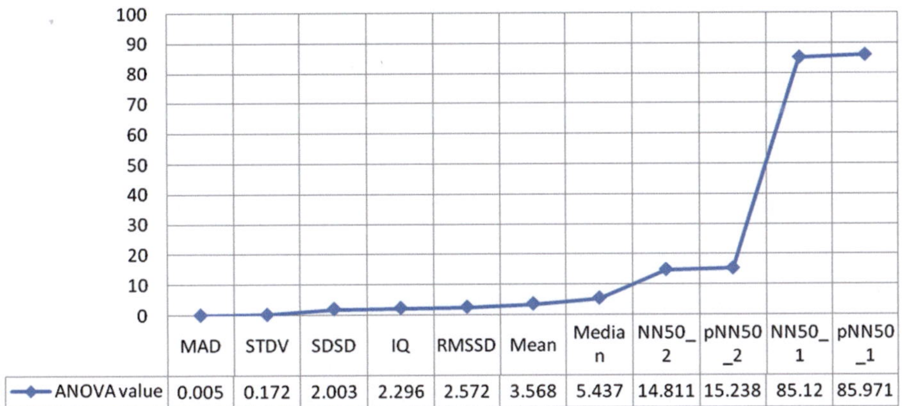

Fig. 5.5 Values of ANOVA test for features set

were extracted from each ECG minute. In order to determine the discriminative power of each feature, ANOVA [31] statistical tests were adopted. ANOVA is a statistical method that stands for analysis of variance. It is used to test the degree to which two or more groups vary or differ in an experiment. For each derived feature, ANOVA value was computed in the objective of identifying the significant ones. The lowest ANOVA value is the highest contribution for the feature. Figure 5.5 presents the ANOVA test value for the features set.

After applying ANOVA test to features vector, it was induced that NN50_1, NN50_2, pNN50_1, and pNN50_2 are the less relevant features and do not contribute highly in the classification results; so they are eliminated from the features set to have seven features instead of 11.

5.3.5 Model Preparation

It is a good practice to prepare data before modeling. An effective data preparation scheme for tabular data when building classification models is standardization. This is where the data is rescaled such that the mean value for each attribute is 0 and the standard deviation is 1. This preserves Gaussian and Gaussian-like distributions while normalizing the central tendencies for each feature.

Classification models will be evaluated using stratified tenfold cross validation. This is a resampling technique that will provide an estimate of the performance of the model. It does this by splitting the data into ten parts, training the model on all parts except one which is held out as a test set to evaluate the performance of the model. This process is repeated ten times, and the average score across all constructed models is used as a robust estimate of performance. It is stratified, meaning that it will look at the output values and attempt to balance the number of instances that belong to each class in the ten splits of the data.

Pipeline utility is also applied in the model preparation stage. The pipeline is a wrapper that executes one or more models within a pass of the cross validation procedure. The goal of using pipeline is to ensure that all of the steps in the pipeline are constrained to the data available for the evaluation, such as the training dataset or each fold of the cross validation procedure. To apply pipeline to our model, the standardization procedure is not only performed on the entire dataset; it is also applied on the training data within the pass of a cross validation run and used to prepare the "unseen" test fold. This makes standardization a step in model preparation in the cross validation process, and it prevents the algorithm having knowledge of "unseen" data during evaluation.

5.3.6   Model Classification

In the classification process, the extracted and selected features have to be fed into the training model to classify each minute of ECG data. In this work, two approaches were proposed for model training process. The first approach is based on the concept of deep learning, and the other one is done using the traditional classification algorithms particularly logistic regression, KNN, SVM, and Naïve Bayes models. In the next subsections, a detail description of the proposed approaches is provided.

5.3.6.1   Deep Learning Approach

Deep learning has shown outstanding results in many applications such as image classification [32], object recognition [33], face recognition [34], and time series data [21, 35].

In our context, the proposed deep learning model passes two phases. In the first phase, a baseline deep neural network (B-DNN) model is proposed. This model is mainly used for the stage of minute-based classification. While in the second phase, a hybrid model is designed by the fusion of deep neural network model and decision tree model. This hybrid model is used for the stage of minute-class-based classification.

Keras [36] library was used for building the proposed deep models. Keras is a highly modular neural networks library, written in Python and capable of running on top of either TensorFlow [37] or Theano [38]. Particularly, scikit-learn package is used to evaluate the model using stratified k-fold cross validation. More details about the proposed deep models are provided in the following subsections.

Phase 1: Baseline deep neural network model (B-DNN)

The proposed baseline deep neural network (B-DNN) model has two fully connected hidden layers. A neural network topology with more layers offers more opportunity for the network to extract key features and recombine them in useful nonlinear ways.

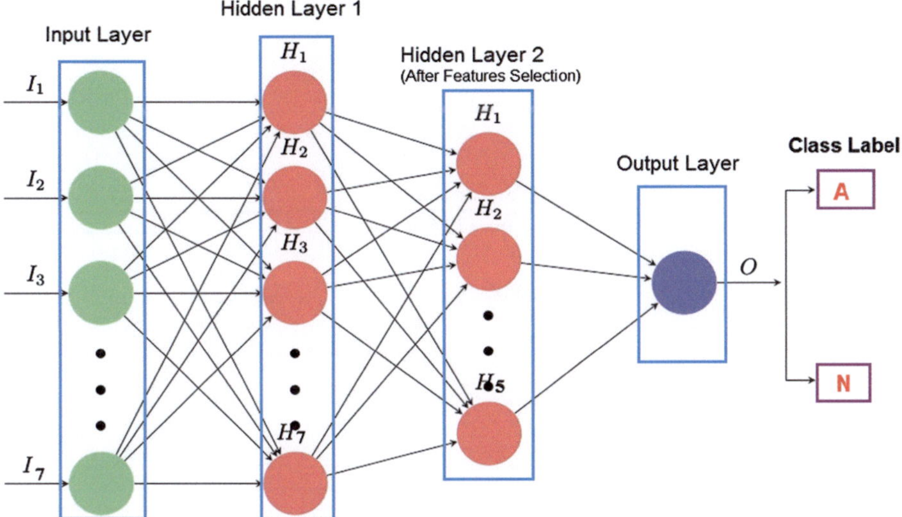

Fig. 5.6 DNN model architecture

The first hidden layer has the same number of neurons as input variables (seven neurons), while the second one was added to force a type of feature extraction by the network by restricting the representational space, since it takes an input of seven neurons (same number as of selected features) and reduces it to five (a new representation of the input features). This will put pressure on the network during training to pick out the most important structure in the input data to model.

Figure 5.6 presents the architecture of the proposed B-DNN model. It shows that the model consists of four layers; the first is the input layer with seven neurons (same number as selected features), the second in the first hidden layer with seven neurons which in turn passes the values to the second hidden layer that squeezes the representational space of the network to have five neurons that is then fed to the output layer which have one neuron that presents the prediction result (apnea or non-apnea).

In B-DNN, the weights are initialized using a small Gaussian random number, and the rectifier activation function is used as a transfer function for the weights to the net input value, which is then fed to the output layer that uses the sigmoid activation function in order to produce a probability output in the range of 0–1 that can easily and automatically be converted to crisp class values. The logarithmic loss function is used during training, which is considered the preferred loss function for binary classification problems. The model also uses the efficient Adam optimization algorithm for gradient descent. Figure 5.7 shows the structure of the B-DNN model.

As aforementioned, this paper proposes a scheme for minute-based classification and minute-class-based classification of sleep apnea. This B-DNN model is used for achieving the first phase, which is minute-based classification.

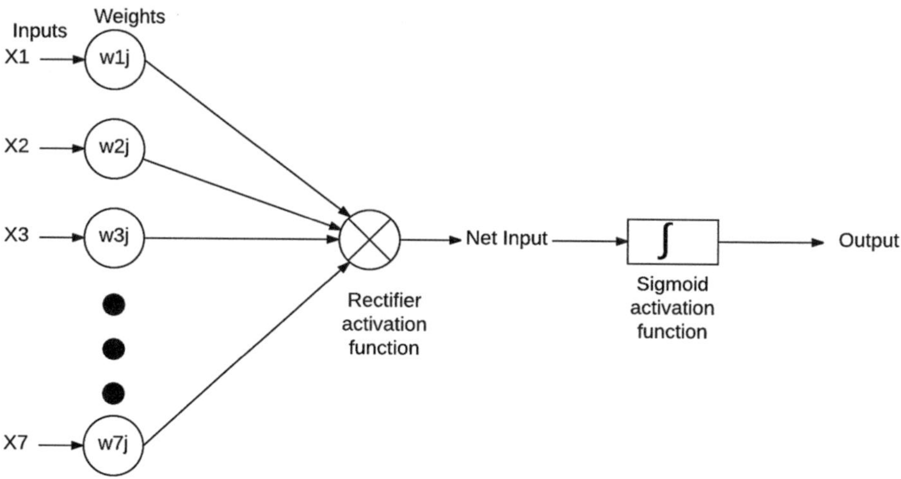

Fig. 5.7 Structure of B-DNN model

Phase 2: Hybrid deep neural network and decision tree model (DNN-DT)

This model is a hybrid algorithm that combines the deep neural network (DNN) classifier with the decision tree classifier. The output of the first phase that is performed using B-DNN model (classified minutes as apnea on non-apnea) is fed into a decision tree model in order to perform class identification (class A, B, or C). Totally, the result is used for the fully minute-class-based classification phase. Figure 5.8 shows the architecture of the proposed DNN-DT.

5.3.6.2 Traditional Classification Algorithms

In addition to our proposed classifier, we applied our extracted and selected features to traditional classifiers being used previously in the literature for the same dataset and compared the results. The explored classifiers are logistic regression, KNN, SVM, and Naïve Bayes classifier. Orange data mining toolset [39] was used to simulate the traditional classifiers and compare results.

5.4 Results

The performance of the proposed classification models is evaluated for both minute-based classification and minute-class-based classification of sleep apnea.

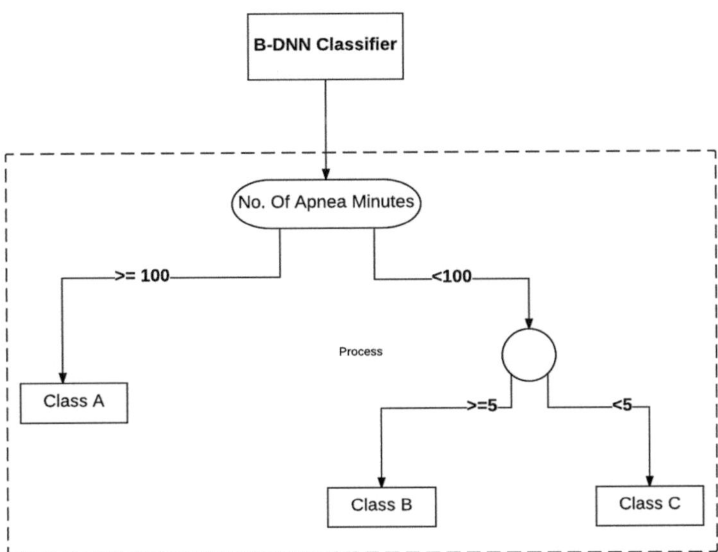

Fig. 5.8 DNN-DT classifier architecture

5.4.1 Features Importance

Features importance is a function that ranks features according to their significance in predicting the target variable of the classification process. Features with higher values contribute the most in the prediction, while features with values near zero do not have high implication on the prediction results. Attributes ranked at zero or less do not contribute to the prediction and should probably be removed from the data [40].

Figure 5.9 presents the results of features importance. The results show that mean absolute deviation (MAD) and standard deviation (STDV) are very important features and have the most effect on whether the minute is classified as apnea on non-apnea minute, while "median" has the less effect on the prediction result.

5.4.2 Minute-Based Classification

Since only the training set (35 ECG recording) of PhysioNet Apnea_ECG dataset contains minute-wise apnea annotations, given the necessity of annotated test data to evaluate the classifier's performance, we were forced to use only these 35 recordings in the experiment. As aforementioned, we evaluated our approach using tenfold cross validation technique.

The performance of the proposed classifier is compared to those of the state-of-the-art classifiers employing Apnea_ECG dataset at the same statistical features.

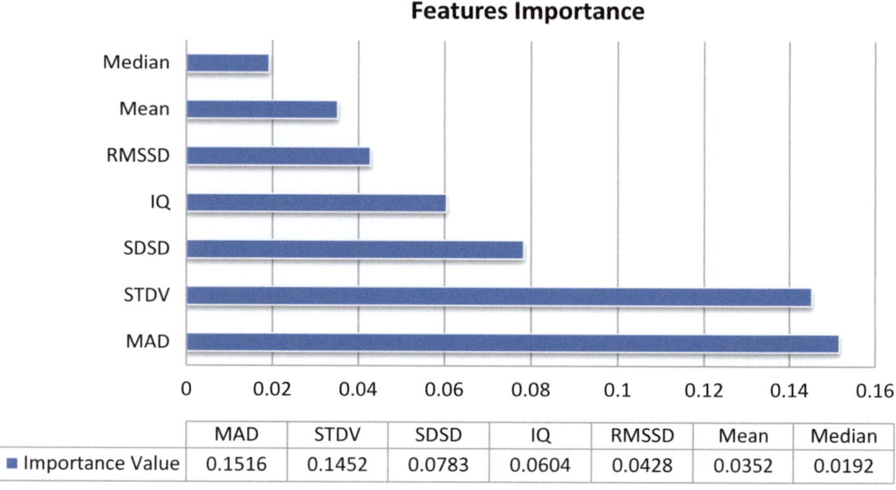

	MAD	STDV	SDSD	IQ	RMSSD	Mean	Median
■ Importance Value	0.1516	0.1452	0.0783	0.0604	0.0428	0.0352	0.0192

Fig. 5.9 Features importance

Table 5.1 A summary of various classifiers performance for minute-based classification

Classifier	# Features = 11					# Features = 7				
	CA	Prec.	Rec.	Sens.	Spec.	CA	Prec.	Rec.	Sens.	Spec.
Logistic regression	60.3	67.3	94.2	94.2	26.4	67.9	66.9	94.9	94.9	23.9
KNN	76.8	80.0	83.2	83.2	66.4	80.5	84.8	83.3	83.3	76.0
SVM	52.3	64.6	50.2	97.1	5.6	62.0	62.3	97.1	97.1	5.6
Naive Bays	63.2	66.0	57.8	57.8	71.8	67.2	76.0	61.5	61.5	68.7
Proposed B-DNN	79.0	80.0	79.0	79.7	77.7	92.7	95.3	92.8	92.8	92.6

This comparison is presented in Table 5.1. It is clear that the proposed B-DNN emerges as the classifier with the highest performance.

Figure 5.10 also shows the performance of all the classifiers before and after features selection. We can observe that in general, classifiers performed better after selecting the top seven relevant features from the whole number of features (11 features). Figure 5.11 also supports these results and presents the classification accuracy for each classifier before and after features selection. The results show that the compact feature subset has good effect on the accuracy of the classification system. Specifically, the accuracy of logistic regression is increased by 7.6%, while that of SVM is increased by more than 10%; accuracy of KNN and Naïve Bayes is increased by about 4.0%. The proposed DNN did the best and achieved a 13.7% increase in accuracy.

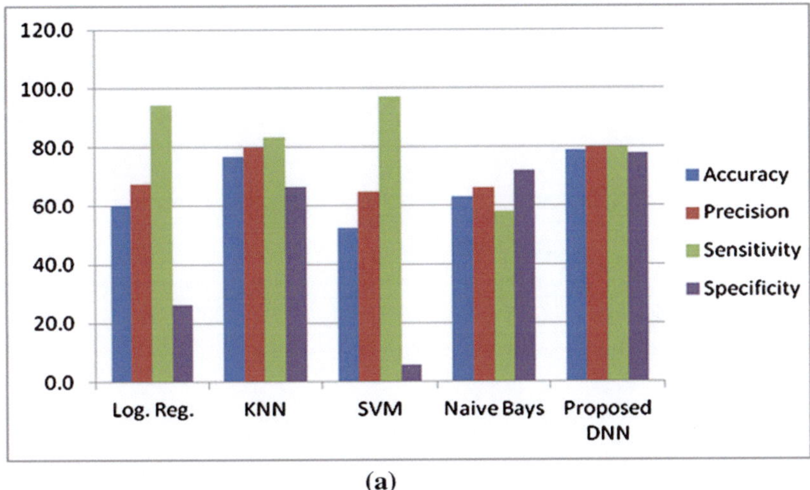

(a)

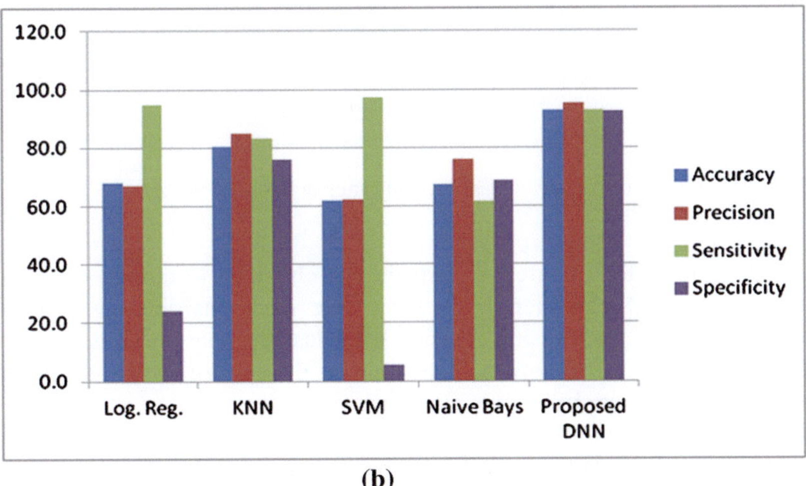

(b)

Fig. 5.10 Classification performance obtained: (**a**) before features selection and (**b**) after features selection

5.4.3 *Minute-Class-Based Classification*

The second phase of the proposed approach is both detection of apnea class and quantification of apnea minutes. The number of the classified minutes for each recording is used to determine whether a patient recording belongs to class A, B, or C, unlike start-of-art methods that were able to classify only two classes instead of three. As mentioned before, the PhysioNet database for the training set contains 20 recordings of class A, 5 of class B, and 10 of class B.

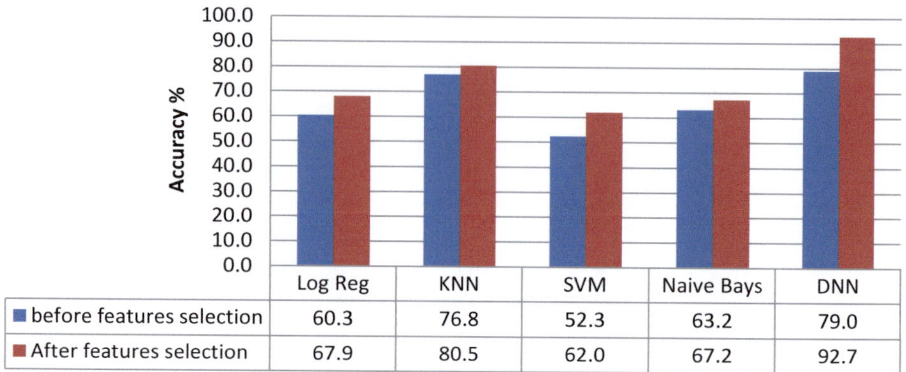

	Log Reg	KNN	SVM	Naive Bays	DNN
before features selection	60.3	76.8	52.3	63.2	79.0
After features selection	67.9	80.5	62.0	67.2	92.7

Fig. 5.11 Accuracy comparison of classifiers performance before and after features selection

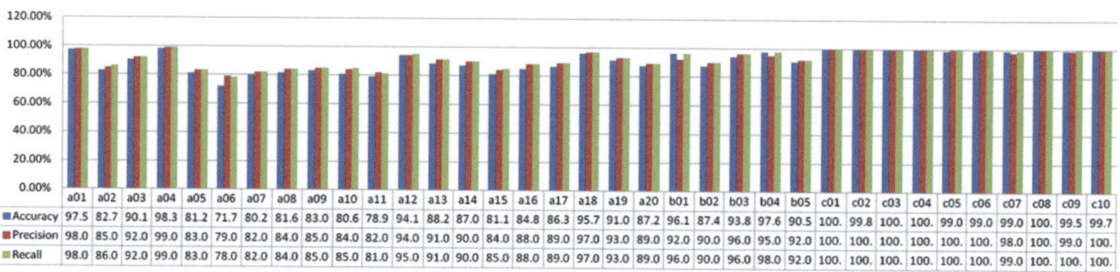

	a01	a02	a03	a04	a05	a06	a07	a08	a09	a10	a11	a12	a13	a14	a15	a16	a17	a18	a19	a20	b01	b02	b03	b04	b05	c01	c02	c03	c04	c05	c06	c07	c08	c09	c10
Accuracy	97.5	82.7	90.1	98.3	81.2	71.7	80.2	81.6	83.0	80.6	78.9	94.1	88.2	87.0	81.1	84.8	86.3	95.7	91.0	87.2	96.1	87.4	93.8	97.6	90.5	100.	99.8	100.	100.	99.0	99.0	99.0	100.	99.5	99.7
Precision	98.0	85.0	92.0	99.0	83.0	79.0	82.0	84.0	85.0	84.0	82.0	94.0	91.0	90.0	84.0	88.0	89.0	97.0	93.0	89.0	92.0	90.0	96.0	95.0	92.0	100.	100.	100.	100.	100.	100.	98.0	100.	99.0	100.
Recall	98.0	86.0	92.0	99.0	83.0	78.0	82.0	84.0	85.0	85.0	81.0	95.0	91.0	90.0	85.0	88.0	89.0	97.0	93.0	89.0	96.0	90.0	96.0	98.0	92.0	100.	100.	100.	100.	100.	100.	99.0	100.	100.	100.

Fig. 5.12 Performance of the minute-based classification at DNN-DT scheme

In more details, the minutes of each files is classified to apnea or non-apnea minute using D-BNN classifier. At the same time, each recording file is classified to its corresponding class (using decision tree classifier) based on the number of classified apnea minutes. Figure 5.12 summarizes the performance of the proposed DNN scheme for minute-by-minute classification.

The performance of the DNN-DT scheme for class-based classification is presented in Fig. 5.13. Hybrid scheme performed well in all performance metrics. Figure 5.14 views the confusion matrix of the scheme. It is also clear that the most misclassified classes are from class B which is a misleading class as induced from other schemes.

5.5 Conclusion

AI and its subbranches, namely, machine learning, deep learning, neural networks, etc., are all algorithmic methods that work on huge amount of data to produce insightful results out of it. This huge amount of data needs to be stored in a format, and most of the times, this format is not our typical SQL database, rather a very unstructured and loosely held information packets.

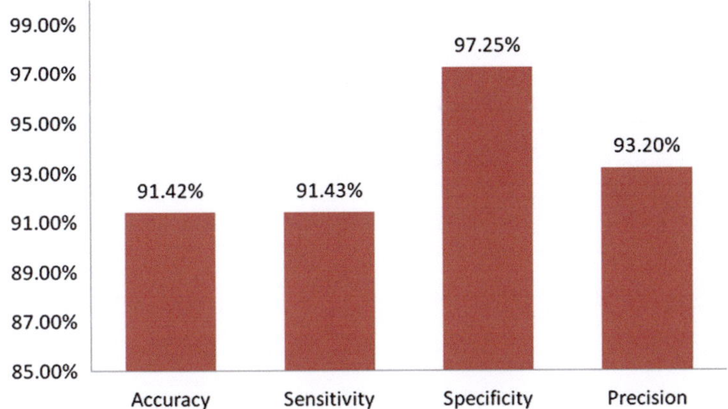

Fig. 5.13 Performance of the class-based classification using DNN-DT scheme

Fig. 5.14 Confusion matrix for the class-based classification using DNN-DT scheme

	A	B	C	Σ
A	19	1	0	20
B	0	3	2	5
C	0	0	10	10
Σ				35

Big Data is a methodology to store very large quantities of data in an unstructured format, and this is exactly what is needed to implement rich intelligent algorithms. So we need a lot of data, a smart algorithm, and a powerful processor which combine to produce data models. So kind of yeah, Big Data or any other practice of large data handling will be used with AI.

Moreover, this plays out in the healthcare; treatment effectiveness can be more quickly determined.

In this chapter, we have proposed a hybrid approach that includes deep neural networks and decision trees for detection and quantification of sleep apnea using features of ECG signals. Statistical features were extracted from the RR interval and serve as training and testing data for the applied classifiers. The proposed approach treated, with novelty, the following points: (1) identifies both apnea classes and detects minute-by-minute classification unlike state-of-the-art methods which either identify apnea class or detect its presence, (2) identifies the three apnea classes (A, B, and C) while other papers only identifies two classes (A and C), and (3) makes a comparative study of the most used classification methods adopted in the literature but using the same features and the same dataset. The experimental results showed that this approach is robust and computationally efficient and clearly outperforms state-of-the-art methods.

Appendix

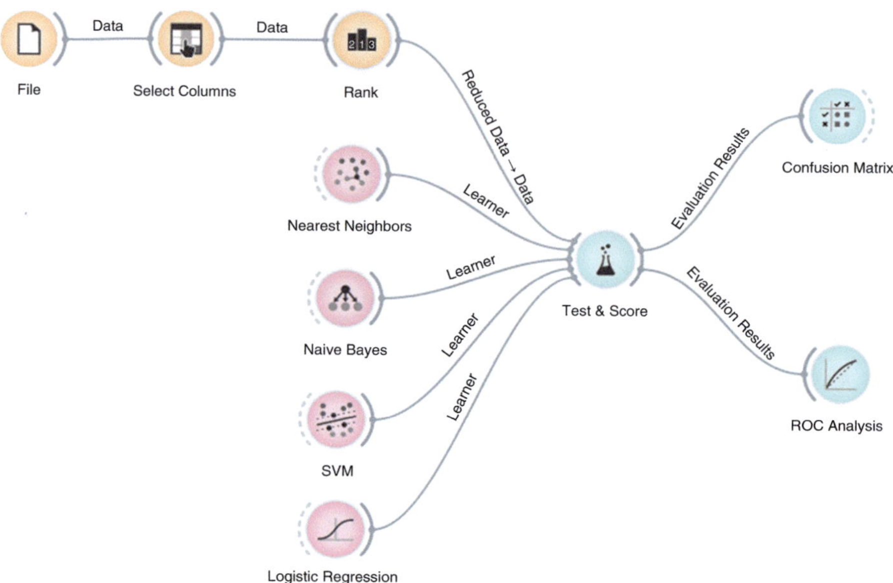

File Name	Actual Class	Predicted Class	Accuracy	Precision	Recall	F1-score	Confusion Matrix
a01	A	A	97.55%	98%	98%	98%	A N Σ A 469 1 470 N 8 11 19 Σ 489
a02	A	A	82.78%	85%	86%	84%	A N Σ A 408 12 420 N 64 44 108 Σ 528
a03	A	A	90.18%	92 %	92%	92 %	A N Σ A 236 10 246 N 31 242 273 Σ 519
a04	A	A	98.38%	99%	99%	99%	A N Σ A 452 1 453 N 3 36 39 492
a05	A	A	81.29%	83%	83%	83%	A N Σ A 242 34 276 N 43 135 178 454

File Name	Actual Class	Predicted Class	Accuracy	Precision	Recall	F1-score	Confusion Matrix
a06	A	A	71.77%	79%	78%	77%	A N Σ A 116 90 206 N 23 281 304 492
a07	A	A	80.24%	82%	82%	82%	A N Σ A 293 29 322 N 62 127 189 511
a08	A	A	81.62%	84%	84%	84%	A N Σ A 146 43 189 N 35 277 312 501
a09	A	A	83.05%	85%	85%	85%	A N Σ A 357 24 381 N 49 65 114 495
a10	A	B	80.65%	84%	85%	84%	A N Σ A 44 56 100 N 20 397 417 517
a11	A	A	78.98%	82%	81%	81%	A N Σ A 159 63 222 N 26 218 244 466
a12	A	A	94.12%	94%	95%	94%	A N Σ A 529 5 534 N 25 18 43 577
a13	A	A	88.27%	91%	91%	91%	A N Σ A 222 22 244 N 25 226 251 495
a14	A	A	87.07%	90%	90%	90%	A N Σ A 367 16 283 N 36 90 126 509
a15	A	A	81.15%	84%	85%	84%	A N Σ A 336 32 368 N 46 96 142 520
a16	A	A	84.86%	88%	88%	88%	A N Σ A 297 23 320 N 35 127 162
a17	A	A	86.35%	89%	89%	89%	A N Σ A 138 20 158 N 33 292 325 483
a18	A	A	95.70%	97%	97%	97%	A N Σ A 433 5 38 N 10 41 51 489
a19	A	A	91.05%	93%	93%	93%	A N Σ A 195 10 205 N 25 272 297 502
a20	A	A	87.21%	89%	89%	89%	A N Σ A 298 17 315 N 39 156 195 510

File Name	Actual Class	Predicted Class	Accuracy	Precision	Recall	F1-score	Confusion Matrix
b01	B	C	96.11%	92%	96%	94%	A: 0, 19, Σ19; N: 0, 468, Σ468; Σ487
b02	B	B	87.44%	90%	90%	90%	A: 70, 23, Σ93; N: 31, 393, Σ424; Σ517
b03	B	B	93.89%	96%	96%	96%	A: 62, 11, Σ73; N: 8, 360, Σ368; Σ441
b04	B	C	97.65%	95%	98%	96%	A: 0, 10, Σ10; N: 0, 416, Σ416; Σ426
b05	B	B	90.51%	92%	92%	91%	A: 25, 32, Σ57; N: 2, 374, Σ376; Σ433
c01	C	C	100%	100%	100%	100%	A: 0, 0, Σ0; N: 0, 478, Σ478; Σ478
c02	C	C	99.80%	100%	100%	100%	A: 0, 1, Σ1; N: 0, 493, Σ493; Σ494
c03	C	C	100%	100%	100%	100%	A: 0, 0, Σ0; N: 0, 454, Σ454; Σ454
c04	C	C	100%	100%	100%	100%	A: 0, 0, Σ0; N: 0, 476, Σ476; Σ476
c05	C	C	99.37	100%	100%	100%	A: 2, 1, Σ3; N: 0, 463, Σ463; Σ466
c06	C	C	99.79	100%	100%	100%	A: 0, 1, Σ1; N: 0, 467, Σ467; Σ468
c07	C	C	99.09%	98%	99%	99%	A: 0, 4, Σ4; N: 0, 425, Σ425; Σ429
c08	C	C	100%	100%	100%	100%	A: 0, 0, Σ0; N: 0, 513, Σ513; Σ513
c09	C	C	99.57%	99%	100%	99%	A: 0, 2, Σ2; N: 0, 455, Σ455; Σ457
c10	C	C	99.77%	100%	100%	100%	A: 0, 1, Σ1; N: 0, 424, Σ424; Σ425

References

1. Derrer, D. (2014, September). *WebMD medical reference*. [Online]. http://www.webmd.com/
2. Caples, S. M. (2007). Sleep-disordered breathing and cardiovascular risk. *Sleep, 30*(3), 291–303.
3. Morgenthaler, T., Kagramanov, V., Hanak, V., & Decker, P. (2006). Complex sleep apnea syndrome: Is it a unique clinical syndrome? *Pub Med Center, 29*(09), 1203–1209.
4. Chazal, P., Penzel, T., & Heneghan, C. (2004, August). Automated Detection of Obstructive Sleep Apnoea at Different Time Scales Using the Electrocardiogram. *Institute of Physics Publishing, 25*(4), 967–983.
5. (2012, January) *Detecting and quantifying apnea based on the ECG*. [Online]. https://www.physionet.org
6. De Chazal, P., et al. (2000). Automatic classification of sleep apnea epochs using the electrocardiogram. *Computers in Cardiology, 27*, 745–748.
7. Jarvis, M., & Mitra, P. (2000). Apnea patients characterized by 0.02 Hz peak in the multitaper spectrogram of electrocardiogram signals. *Computers in Cardiology, 27*, 769–772.
8. Mcnames, J., & Fraser, A. (2000). Obstructive sleep apnea classification based on spectrogram patterns in the electrocardiogram. *Computers in Cardiology, 27*, 749–752.
9. Mietus, J., Peng, C., Ivanov, P., & Goldberger, A. (2000). Detection of obstructive sleep apnea from cardiac interbeat interval time series. *Computers in Cardiology, 27*, 753–756.
10. Schrader, M., Zywietz, C., Einem, V., Widiger, B., & Joseph, G. (2000). Detection of sleep apnea in single channel ECGs from the PhysioNet data base. *Computers in Cardiology, 27*, 263–266.
11. Raymond, B., Cayton, R., Bates, R., & Chappell, M. (2000). Screening for obstructive sleep apnoea based on the electrocardiogram – The computers in cardiology challenge. *Computers in Cardiology, 27*, 267–270.
12. A Khandoker, C Karmakar, and M Palaniswami, "Automated recognition of patients with obstructive sleep apnoea using wavelet-based features of electrocardiogram recordings," *Computers in Biology and Medicine*, vol. 39, no. 3, pp. 88–96, 2009.
13. Xie, B., & Minn, H. (2012). Real-time sleep apnea detection by classifier combination. *Information Technology in Biomedicine, 16*(3), 469–477.
14. Manrique, Q, Hernandez, A, Gonzalez, T, Pallester, F, & Dominquez, C. (2009). Detection of obstructive sleep apnea in ECG recordings using time-frequency distributions and dynamic features. In *IEEE International Conference on Engineering in Medicine and Biology Society(EMBS 2009)*, pp. 5559–5562.
15. Mendez, M., et al.. (2007). Detection of sleep apnea from surface ECG based on features extracted by an autoregressive model. In *IEEE International Conference on Engineering in Medicine and Biology Society (EMBS 2007)*, pp. 6105–6108.
16. Almazaydeh, L., Elleithy, K.H., & Faezipour, M. (2012). Obstructive sleep apnea detection. In *IEEE International Conference on Engineering in Medicine and Biology Society (EMBS 2012)*.
17. Babaeizadeh, S., White, D., Pittman, S., & Zhou, S. (2010). Automatic detection and quantification of sleep apnea using heart rate variability. *Journal of Electrocardiology, 43*, 535–541.
18. Rachim, V., Li, G., & Chung, W. (2014). Sleep apnea classification using ECG-signal wavelet-PCA features. *Bio-Medical Materials and Engineering, 24*, 2875–2882.
19. Zeiler M.D., Fergus R. (2014) Visualizing and Understanding Convolutional Networks. In: Fleet D., Pajdla T., Schiele B., Tuytelaars T. (eds) Computer Vision – ECCV 2014. ECCV 2014. Lecture Notes in Computer Science, vol 8689. Springer, Cham.
20. Simonyan K., Vedaldi A, Zisserman A. (2014) Deep Inside Convolutional Networks: Visualising Image Classification Models and Saliency Maps. Computer Vision and Pattern Recognition https://arxiv.org/abs/1312.6034v2.
21. Hinton, G., et al. (2012). Deep neural networks for acoustic modeling in speech recognition: The shared views of four research groups. *IEEE Signal Processing Magazine, 29*, 82–97.

22. Brébisson, A. D., & Montana, G. (2015). Deep neural networks for anatomical brain segmentation. In 2015 *IEEE Conference on Computer Vision and Pattern Recognition Workshops (CVPRW)* (pp. 20–28).
23. Wang L, et al. (2011) Growth propagation of yeast in linear arrays of microfluidic chambers over many generations. Biomicrofluidics 5(4):44118-441189.
24. Fukushima, K., & Miyake, S. (1982). Neocognitron: A new algorithm for pattern recognition tolerant of deformations and shifts in position. *Pattern Recognition, 15*(6), 455–469.
25. Bengio, Y., Courville, A., & Vincent, P. (2013). Representation Learning: A review and new perspectives. *IEEE Transactions on Pattern Analysis and Machine Intelligence, 35*(8), 1798–1828.
26. Goldberger, A., et al. (2000). PhysioBank, PhysioToolkit, and PhysioNet: Components of a new research resource for complex physiologic signals. *Circulation, 101*(23), e215–e220.
27. MedicineNet. (2016, September) *Definition of QRS complex.* [Online]. http://www.medicinenet.com/script/main/art.asp?articlekey=5160
28. Thuraisingham, R. (2006). Preprocessing RR interval time series for heart rate variability analysis and estimates of standard deviation of RR intervals. *Computer Methods and Programs in Biomedicine, 83*(1), 78–82.
29. (2015, July) *The WFDB Software Package.* [Online]. https://www.physionet.org/physiotools/wfdb.shtml
30. Kaguara, A., Myoung Nam, K., & Reddy, S. (2014, December). *A deep neural network classifier for diagnosing sleep apnea from ECG data on smartphones and small embedded systems.* Thesis.
31. (2013). *Statistics solutions.* [Online]. http://www.statisticssolutions.com/manova-analysis-anova/
32. Hayat, M., Bennamoun, M., & An, S. (2015). Deep reconstruction models for image set classification. *IEEE Transactions on Pattern Analysis and Machine Intelligence, 37*, 713–727.
33. Bai, J., Wu, Y., Zhang, J., & Chen, F. (2015). Subset based deep learning for RGB-D object recognition. *Neurocomputing, 165*, 280–292.
34. Huang, Z., Wang, R., Shan, S., & Chen, X. (2015). Face recognition on large-scale video in the wild with hybrid Euclidean-and-Riemannian metric learning. *Pattern Recognition, 48*, 3113–3124.
35. Deng, J., Zhang, Z., Eyben, F., & Schuller, B. (2014). Autoencoder-based unsupervised domain adaptation for speech emotion recognition. *IEEE Signal Processing Letters, 21*, 1068–1072.
36. Keras Documentation. [Online]. https://keras.io/
37. (2016). TensorFlow. [Online]. https://www.tensorflow.org/
38. LISA Lab. (2016, August). Theano. [Online]. http://www.deeplearning.net/software/theano/
39. (2016, August). *Orange data mining.* [Online]. http://orange.biolab.si/
40. Geoffrey, E., Hinton, N. S., Krizhevsky, A., Sutskever, I., & Salakhutdinov, R. R. (2012). Improving neural networks by preventing co-adaptation of feature detectors. *Neural and Evolutionary Computing.* https://arxiv.org/abs/1207.0580v1

Chapter 6
A Study of Data Classification and Selection Techniques to Diagnose Headache Patients

Ahmed J. Aljaaf, Conor Mallucci, Dhiya Al-Jumeily, Abir Hussain, Mohamed Alloghani, and Jamila Mustafina

6.1 Introduction

The application of big data solutions, enabled by data science approaches and intelligent system technologies, has already delivered transformative impacts in the health domain. Firstly, the expansion in basic science frameworks toward data-intensive processes, using connected and collaborative operational models, has enabled advances in health critical areas including genomics, neuroscience, pharmaceutical development, systems biology, bioinformatics and others. Prominent related 'Big Science' projects include the Human Brain Project in Europe, the Blue Brain Project in Switzerland, the Brain Activity Map in the USA and the BRAIN initiative that is also based in the USA. Such ambitious large-scale projects, which make extensive use of data science approaches, such as data mining, serve to

A. J. Aljaaf (✉)
Faculty of Engineering and Technology, Liverpool John Moores University, Liverpool, UK

Center of Information Technology, University of Anbar, Ramadi, Iraq
e-mail: A.J.Kaky@ljmu.ac.uk

C. Mallucci
Alder Hey Children's NHS foundation trust, Liverpool, UK

D. Al-Jumeily · A. Hussain
Faculty of Engineering and Technology, Liverpool John Moores University, Liverpool, UK
e-mail: D.Aljumeily@ljmu.ac.uk; A.Hussain@ljmu.ac.uk

M. Alloghani
Abu Dhabi Health Services Company (SEHA), Abu Dhabi, UAE
e-mail: M.AlLawghani@2014.ljmu.ac.uk

J. Mustafina
Kazan Federal University, Kazan, Russia
e-mail: DNMustafina@kpfu.ru

© Springer International Publishing AG, part of Springer Nature 2018
M. M. Alani et al. (eds.), *Applications of Big Data Analytics*,
https://doi.org/10.1007/978-3-319-76472-6_6

deliver fundamental insights into human biology and brain function, unlocking new therapeutic solutions. Furthermore, big data and intelligent systems approaches have been applied to the growing space of patient medical data to derive new solutions in disease prediction, patient monitoring, diagnosis, pre-surgical evaluation and world disease burden analysis, among other problem domains. Such a space of intelligent solutions opens up new emerging paradigms in healthcare including P4 medicine (predictive, personalized, preventive and participatory), enabling an emphasis on wellness as opposed to disease.

The development of diagnostic models or clinical decision support models (*CDSMs*) to aid in the diagnosis of primary headache disorders has become an interesting research topic, especially after the launch of the International Headache Society IHS clinical criteria for the classification of headache [1]. A range of studies or diagnostic models have been proposed or already developed to aid headache specialists in making decisions with respect to the diagnosis of headaches. Many others were restricted for patients' usage such as an application to enable patients in keeping track of their conditions and treatments or applications to get recommendations from health professionals. This chapter reviews the literature to investigate recent expert systems or *CDSMs* that target the diagnosis of primary headache disorders. This chapter also analyses the core concept of these *CDSMs* to explore their advantages and drawbacks, which enable us to initialize a new pathway toward robust diagnostic model that overcomes current challenges.

6.2   Intelligent-Driven Models

This section reviews the most recent diagnostic models or studies that have been published over the last decade. We have searched a wide range of scientific libraries and resources such as PubMed, Google Scholar, Springer Link, IEEE Xplore digital library and many other scientific journals. The search strategy included the following key words: 'expert system, intelligent diagnosis, primary headaches, migraine, cluster and tension-type headache'.

As presented in Fig. 6.1, we came up with 22 studies employing intelligent approaches based on our search strategy; 11 out of 22 studies were out of the scope of this research for the following reasons: the first 4 studies were not related to the diagnosis of primary headaches, while knowledge acquisition was not clearly mentioned for 3 studies. We ended up with 15 most relevant studies; 4 of them employed machine-learning approaches, which are also not considered in this study. Therefore, 11 selected studies that build around knowledge-based approach will be reviewed, discussed and evacuated with respect to the diagnosis of primary headache. The 11 relevant studies have proposed different clinical decision support models (CDSM) with the aim of improving the diagnosis of primary headache disorders.

Before starting to assess these studies, we need to have a deep understanding for the driven force or reasons behind such diagnostic models. In general, headache

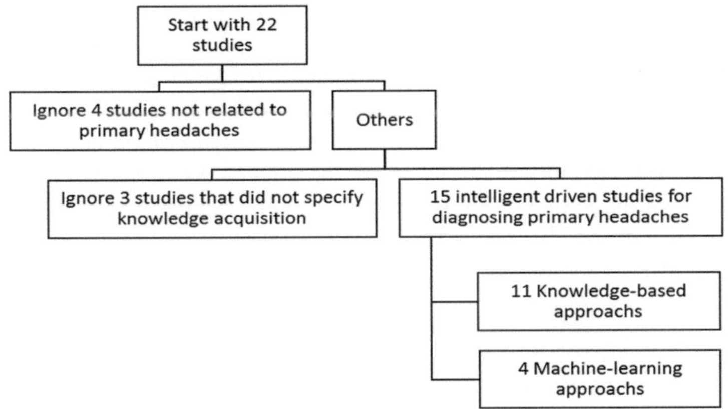

Fig. 6.1 Block diagram of search strategy

diagnosis made by specialists depends on clinical guidance, medical evidence, instructions and principles derived from medical science. Ideally, *CDSMs* should improve the use of knowledge to support those specialists in making accurate decisions and therefore enhancing the quality of care being delivered to headache patients. Although *CDSMs* have a potential to improve decision-making, handling large amount of information and analysing real-time data or patient history, however, the use of *CDSMs* are not yet widespread in clinics or hospitals. This may be due to the fact that the majority of such models or system are developed apart from healthcare professionals and there is lack of criteria for a proper use of intelligent methods in these *CDSMs* [2]. In 2015, we have established a simple yet powerful yes/no inquiry method to help in identifying the most proper use of intelligent approaches in the *CDSMs* based on observations of previous studies [3].

In this study, we investigate how the proposed knowledge-based *CDSMs* have been developed and how the knowledge acquisition stage has been processed for each study. First, we grouped the relevant *CDSMs* into two main categories: *CDSMs* that employed machine learning and *CDSMs* that followed knowledge-based system style. The main difference between these two groups is that machine-learning approaches are able to learn the important features of a given headache data set in order to make predictions about new headache cases [4]. In other words, machine-learning-based *CDSMs* can learn and gain knowledge from pre-diagnosed headaches and/or through identifying patterns in the medical data. Machine-learning-based *CDSMs* will be covered in the upcoming studies.

Conversely, the core concept of knowledge-based *CDSMs* are designed and structured around the logic of *IF-THEN* statements, in which clinical guidelines such as IHS criteria or experts' knowledge are formed into rules and expressed by a computer engineer as a set of *IF-THEN-ELSE* statements. This usually includes a significant amount of information regarding the types of headache together with their signs and symptoms. Once the patient data are inputted, the inference engine examines the data against these IF-THEN statements to limit the outcome response. A simple example of using knowledge-based core model is presented in [5], where the proposed decision support model includes a probable list of haematological diseases combined with their symptoms. Patient information such as age, gender, altitude and pregnancy period in addition to the complete blood count (CBC) test result will be inputted to this model. Then the inference engine will suggest a list of probable haematological diseases based on these inputs. Although it is unable to provide an ultimate diagnosis, however, it is a good start for further and more disease-specific tests to confirm the diagnosis.

6.3   Knowledge-Based Headache *CDSMs*

The majority of works that have been done to improve the diagnosis of primary headaches followed knowledge-based systems style, in which the knowledge derived from clinical guidelines, i.e. IHS criteria, and formulated as a set of diagnostic roles by a human expert. This style is commonly known as expert system or rule-based system. Expert system-based headache solution (ESHS) was proposed by Hasan et al. [6] to diagnose different types of headache including migraine and cluster headache. ESHS includes a set of key questions derived from neurology experts to help other doctors when diagnosing patients with headache. When symptoms entered in accordance with these questions, ESHS then would help in detecting the type of headache and generate prescriptions. The expert system uses very simple yes/no questions derived from expert's knowledge without clarifying who are the experts and show their affiliations and experiences. This major drawback of ESHS raises concerns about the validity of diagnostic questions as well as the feasibility of applying such CDSM.

Al-Hajji [7] has developed rule-based CDSM to diagnose more than ten types of neurological diseases including migraine and cluster headache. In this CDSM, knowledge has been obtained from different sources such as domain experts, specialized databases, books and a few electronic websites. A list of neurological diseases has been stored in a table, and approximately 70 related symptoms were also stored in another table. Then, a combination between each neurological disease and its most related symptoms has been derived. In fact, the diagnosis of many neurological diseases, such as Alzheimer's, Parkinson's and epilepsy, in addition to migraine and cluster headache, can be challenging even for neurology specialists themselves. It is a wide range of diseases that generally have shared symptoms and various diagnostic procedures. For example, brain imaging can play a vital role in the diagnosis of Alzheimer's or the early detection of Parkinson's disease. Moreover, there was no clear adoption of IHS criteria with respect to the diagnosis of migraine and cluster headache. Therefore, using a very simple link between each neurological disease and its symptoms cannot be seen as an effective clinical DSS and would bear a large error rate.

A computerized headache guideline method was proposed by Yin et al. [8] to assist general practitioners in primary hospitals with the diagnostic of primary headaches such as migraine, tension-type headache and cluster headache. The main aim was to develop a system to counteract the complexity of the second version of IHS criteria. Authors pass through three main steps to develop their CDSM. A clinical specialist summarizes the diagnostic guidelines of IHS and expresses them as a flowchart in the first step. Then, a knowledge engineer establishes a computerized model for headache knowledge representation based on these flowcharts. Finally, the knowledge representation model is translated into a series of conditional rules, which are used by the inference engine. This CDSM was evaluated by 282 previously diagnosed headache cases obtained from a Chinese hospital.

In 2014, Dong and his colleagues have developed a guideline-based CDSM for headache diagnosis [9]. They have followed the same procedure presented in [8] for knowledge acquisition, but using the third version of IHS criteria, and validated their system by 543 data sheet of patients with headache obtained from the International Headache Centre at the Chinese PLA General Hospital, Beijing, China. The main difference between this guideline-based CDSM and the guideline-based CDSM developed by Yin in [8] is that three more types of headache have been added to the library of this DSS including probable migraine, probable tension-type headache, new daily persistent headache and medication overuse headache. As shown in [9], there was some improvement in the diagnosis in comparison with CDSM by Yin in [8].

In 2015, Yin et al. [10] have proposed computer-aided diagnosis method that employs case-based reasoning (CBR) method to differentiate between probable migraine and probable tension-type headache. This CBR CDSM provides recommendations to the general practitioners based on the previously solved cases in the built-in library. This library contains 676 data sheets of patients with probable migraine and probable tension-type headache that were collected by clinical interview. Each data sheet consists of 74 different attributes including patients' information and medical history in addition to headache symptoms derived from the IHS criteria. The authors employ genetic algorithm (GA) to assign weights to these attributes and K-nearest neighbor (KNN) method to measure the similarity between new headache cases and the previous cases in the library.

A hybrid CDSM tool was proposed by Yin et. al [11] using a combination of rule-based and case-based reasoning methods to improve the diagnosis of primary headache disorders such as migraine, tension-type headache and cluster headache. The reasoning modules in this CDSM run independently, the rule-based module is the first diagnostic module, and the case-based module is the second. The diagnostic rules are summarized by a clinical specialist based on the criteria of IHS in the first module, while data sheets of previous headache cases have been used in the second module. The diagnostic procedure starts through applying the first diagnostic module to a new headache case; if headache symptoms are typical and match the existing rules, then a diagnostic decision can be made. Otherwise, the headache case is transferred to the case-based module to search for the most similar previous cases.

The research group in [10] claim that the CBR CDSM shows an improvement with respect to the diagnosis of primary headaches when compared to their previous works [8, 9] built around the guideline-based concept. Although the core concept of [8, 9] and [10] seems to be similar, however, knowledge acquisition methods are completely different. In [8, 9], the specialist derives diagnostic guidelines from IHS criteria, which is then expressed as a set of conditional rules, while [10] uses clinical interviews of patients with headache as a knowledge acquisition stage. The same research group have also proposed a hybrid CDSM in [11], which is a merger of their previous proposals in [8–10].

Many other *CDSMs* have been proposed over the last decade; Simić et al. in [12, 13] have proposed a computer-assisted diagnosis of primary headaches. It is a rule-based fuzzy logic (RBFL) system designed to help physicians when diagnosing patients with primary headaches such as migraine, tension-type headache and cluster headache. This work involves the type of knowledge-based CDSM, in which the criteria of IHS are expressed as a collection of IF-THEN statements.

Eslami et al. in [14] have designed a computerized expert system to help in the diagnosis of primary headache disorders such as migraine, tension-type headache, cluster headache and other trigeminal autonomic cephalalgias. A questionnaire was designed to approach all criteria of primary headache disorders based on the second version of IHS criteria. When a patient starts filling in the questionnaire, the expert system uses a simple human-like algorithmic reasoning to classify the type of headache. Similarly, Maizels and Wolfe in [15] employ a simple human-like branching logic to determine the most appropriate diagnostic questions to ask the patients and then classify the type of headache using modified Silberstein-Lipton criteria and IHS criteria. Maizels and Wolfe implemented their expert system as a web-based tool with an interview section that includes questions about headache characteristics. The modified Silberstein-Lipton criteria are used to classify patient with frequent headache, while IHS criteria are used to diagnose patients with brief headache syndromes.

Zafar et al. in [16] proposed a CDSM to aid physicians in the diagnosis of migraine and other headaches and at the same time to enable patients living in remote areas to have medical check-ups. Zafar implemented his work as a web-based tool, in which information related to primary and secondary headaches are stored in the knowledge base. The inference engine will search this knowledge base to find suitable diagnostic recommendations based on headache characteristics. This proposed system, in fact, is considered as a black box because there is no clear sequence of operations in particular for knowledge acquisition.

The following Table 6.1 summarizes *CDSMs* targeted in this study; it covers studies conducted over the past 8 years. It is obvious that knowledge-based *CDSMs* were built around the classification criteria of IHS. Although some of the *CDSMs* employ questionnaire or use a domain expert, however, it is an indirect use of the IHS classification of headaches. In the next section, we will explain how knowledge will be extracted from clinical guidelines such as the criteria of HIS.

Table 6.1 Knowledge-based diagnostic models

No.	Authors	Year	Knowledge acquisition	Type of headache
1	Al-Hajji [7]	2012	Domain experts	Migraine and cluster headache
2	Hasan et al. [6]	2012	Domain experts	Primary headaches
3	Yin et al. [8]	2013	IHS criteria	Primary headaches
4	Dong et al. [9]	2014	IHS criteria	Primary headaches
5	Yin et al. [10]	2015	Case-based similarity	Probable migraine and probable tension-type headache
6	Yin et al. [11]	2014	Case-based and IHS criteria	Primary headaches
7	Simić et al. [12, 13]	2008	IHS criteria	Primary headaches
8	Eslami et al. [14]	2013	Questionnaire	Primary headaches
9	Maizels and Wolfe [15]	2008	Silberstein-Lipton criteria and IHS criteria	Primary headaches
10	Zafar et al. [16]	2013	Unknown	Primary and secondary headaches

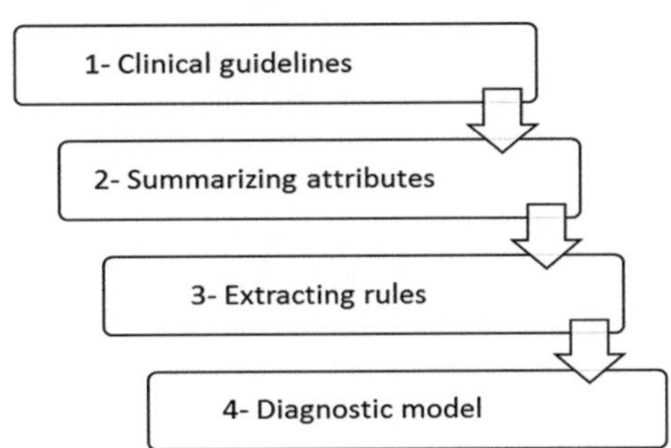

Fig. 6.2 Steps of knowledge acquisition for headache diagnosis

6.4   Knowledge Acquisition

Knowledge acquisition is the process of extracting, analysing and translating knowledge used by human expert when managing or solving problems. It is the major bottleneck in the development of expert systems, in which the process of interpreting knowledge is complex, challenging and usually time-consuming. Knowledge-based headache *CDSMs* followed similar way regarding the knowledge acquisition, where they have used the criteria of IHS for classification of headache disorders as a base for deriving the diagnostic rules. We can summarize the process of knowledge acquisition for headache diagnosis in four steps as shown in Fig. 6.2.

6.4.1 Clinical Guidelines

IHS has established a standardized terminology and consistent operational diagnostic criteria for a wide range of headache disorders [17]. These criteria were drawn up based on an international consensus of headache experts and have been accepted worldwide as a gold standard for headache diagnosis. The IHS uses straightforward diagnostic criteria, which are explicit, unambiguous, accurate and with as little scope for interpretation as possible. In this study, we are focusing on the diagnosis of primary headaches. Primary headache disorders are the most common in the community; they are not related to any underlying medical condition, and the headache itself is the disorder [18]. In contrast, secondary headache disorders occur secondarily to another medical condition, some of which may be life threatening and therefore require quick and accurate diagnosis. Secondary headache is extremely rare and represents less than 1% of the population who experience headaches [19].

Migraine is the commonest debilitating and disabling primary headache disorder. Including both chronic migraine (CM) and episodic migraine (EM) forms, it affects up to 18% of women, less frequently in men [20, 21]. Patients could meet the criteria of migraine without aura by different combinations of features; no single feature is essential to be present. Because two of four pain features are required, therefore, a patient with unilateral, throbbing pain could be eligible to meet the criteria, so does a patient with moderate pain that is aggravated by physical activity. Likewise, only one of two possible related symptom combinations is required. Patients with nausea or vomiting, but without photophobia or phonophobia, meet the conditions, as do patients with photophobia and phonophobia but without nausea or vomiting [17].

Tension-type headache (TTH) is a very common form of primary headache [17], with a lifetime prevalence ranging from 30% to 78% in the general population as shown by several studies [1, 22]. According to the criteria of IHS, the diagnostic criteria for tension-type headache have primarily been designed to differentiate between tension-type headache and migraine [1]. In contrast to migraine, the main pain features of tension-type headache can be represented by the absence of migraine's characteristic features. The pain is mild to moderate and not as severe as in migraine, non-throbbing quality and not aggravated by physical activity. No nausea or vomiting is associated, although no more than one of phonophobia or photophobia [17, 18, 20, 23]. The headache can be unilateral but is commonly generalized. It can be described as pressure or tightness, such as a tight band around the head, and usually arises from or spreads into the neck [19].

Cluster headache (CH) is the commonest form of the TACs. CH predominantly appears in young adulthood as early as the second decade of age, persisting well in life, even in the seventh decade [24]. CH is extremely rare in children, men are also more than three times more likely to be diagnosed with this type of headache, and it is quite often in smokers [17, 25]. CH is usually severe, recurring, but generally briefer than migraine and non-throbbing [3]. The pain is excruciatingly severe, intense, strictly unilateral and variously described as sharp, drilling and stabbing [17]. It is most often located behind one eye and sometimes generalized to a larger area of the head [19]. In general, the pain takes 10–15 min to reach its peak intensity and remains excruciatingly intense for an average of 1 h and usually ranges from 15 to 180 min. Typically, it occurs at the same time every day, most often at night, 1–2 h after sleep [17, 19]. Patients during the attack find it difficult to lie down, because it aggravates the pain and can cause themselves harm through beating their head on the wall or floor until the pain reduces, usually after 30–60 min [17, 19].

6.4.2 Summarizing Attributes

Primary headaches may share certain features; pain is severe for migraine and CH as an example. However, CH varies from migraine primarily in its pattern of occurrence. CH is in briefer episodes over a period of weeks or months. Sometimes, a whole year can pass between two CHs. Migraine usually does not follow this type of pattern. Most of the migraine's features explicitly differentiate this type of headache from TTH and therefore help in a precise diagnosis. Similar to episodic TTH, migraine is a recurrent headache that can last from a couple of hours to a few days. However, while TTH is commonly generalized, migraine pain is mostly unilateral; and while migraine has a pulsating quality with moderate to severe pain, TTH presents as a mild to moderate in intensity and a dull ache or feeling of a tight band around the head [26, 27]. Furthermore, patients with TTH headache are significantly less disabled than patients with migraine or cluster headache [17].

Before starting to summarize headache attributes, we have initialized a comparison table of the three main types of primary headache. Table 6.2 illustrates the distinct as well as the overlapping signs and symptoms of TTH, migraine and cluster headache. From this table we will start to summarize the required headache attributes as a step toward creating the diagnostic rules. So let us consider the following: $D(x, y)$ is the headache duration, where x and y represent minimum and maximum boundaries of headache duration according to IHS criteria. L is the

Table 6.2 Comparison of primary headaches clinical features

	Migraine	TTH	Cluster
Gender ratio (M:F)	1:3	5:4	3:1
Age of onset	15–55 years	25–30 years	28–30 years
Prevalence	18% F – 6% M	30 up to 78%	0.9%
Quality	Throbbing	Non throbbing	Stabbing, sharp
Intensity	Moderate to severe	Mild to moderate	Severe to very severe
Location	Unilateral	Bilateral	Unilateral
Duration of attack	4–72 h	30 min to 7 days	15–180 min
Symptoms	Nausea, vomiting, photophobia, phonophobia	Photophobia, phonophobia	Autonomic dysfunction
Triggers	Physical activity	Stress	Laying down or sleep

headache location with two possible values unilateral or bilateral. Q is the headache quality with three possible values including pulsing pain, non-pulsing pain and stabbing pain quality. T is the headache triggers including three possible factors (i.e. physical activity, stress and laying down). I is headache intensity with four possible values including mild pain, moderate, severe and very severe pain intensity. S represents the symptoms that may accompany pain, which including all symptoms that may accompany the primary headache as mentioned in can be one or more of the following symptoms nausea, vomiting, photophobia, phonophobia or autonomic dysfunction symptoms. A is the presence/absence of aura symptoms with headache, where the headache can be free of aura or altered vision, tingling in the body and weakness of the arms and legs. P represents how many headache days per month and its range from 1 to 30. Finally, G is a patient gender.

6.4.3 Extracting and Formulating Diagnostic Rules

After identifying the clinical guidelines and summarizing the required headache attributes, in this section we will create procedural functions as shown in Fig. 6.3 to help in the classification of primary headaches. As an example, we present a function to diagnose migraine patients in accordance with the criteria of IHS and use the summarized headache attributes. We are initializing migraine function with the migraine constant attributes first, where migraine is characterized by a unilateral pain location and throbbing pain quality and aggravated by routine physical activity. Then we identify the ranges for all other attributes that could vary from patient to another. Finally, we use a switch-case conditional statement to classify the type of migraine. This part can also be formulated using a sequence of IF-THEN statements. For simplicity, we replace attributes of possible options with dummy numbers as shown in the following equations, where L will be represented as 0 when headache is unilateral and 1 when headache is bilateral. Likewise, S will be represented by an integer number, $S = \{s_1, s_2,, s_m\}$, $s_i \in N^*$ and $1 \leq i \geq m$, where m represents the number of the symptom that may associate with headache as mentioned in Table 6.2 and so on.

How knowledge is acquired and expressed from the criteria of IHS and formulated as diagnostic tools for primary headaches is now clear. This CDSM can help non-specialist doctors and general practitioners with respect to the diagnosis of primary headache disorders.

6.5 Summary and Limitations

This study reviews, examines and shows the core concept of the proposed knowledge-based headache diagnostic models influenced by the international classification of headache disorders established by IHS. The development of

```
1.  Start
2.  Constants:
3.      L = 0; // unilateral
4.      Q = 1; // pulsing pain
5.      T = 2; // physical activity;
6.  Variables:
7.      D = random selection (240, 4320); // headache duration calculated in minutes
8.      I = random selection (1, 2); // pain intensity
9.      S = random selection (1, 4); // symptoms
10.     A = random selection (1, 4); // aura symptoms
11.     P = (4, 30); // headache period
12.     G = random selection (0, 1); // patient gender
13.
14. // N is the number of required instances (number of patients record will be generated).

15. For i| = 1 to i <= N Do
16. Begin
17. Switch (A)
18. Begin
19.      Case (A == 1)
20.          If (P <= 14)
21.              Begin
22.                  Headache = episodic migraine without aura;
23.                      Print (D, L, Q, I, T, S, A, P, G, Headache);
24.              End if
25.          Else
26. Begin
27.                  Headache = chronic migraine without aura;
28.                      Print (D, L, Q, I, T, S, A, P, G, Headache);
29.              End else
30.      Break;
31.      Case (A > 1 or A <= 4)
32.          If (P <= 14)
33.              Begin
34.                  Headache = episodic migraine with aura;
35.                      Print (D, L, Q, I, T, S, A, P, G, Headache);
36.              End if
37.          Else
38. Begin
39.                  Headache = chronic migraine with aura;
40.                      Print (D, L, Q, I, T, S, A, P, G, Headache);
41.              End else
42.      Break;
43. Default:
44. Break;
45. End switch
46. End for
47. End start;
```

Fig. 6.3 Procedural migraine classification function

diagnostic models for primary headaches has been widely targeted by researchers rather than secondary headaches due to the following reasons. First of all, primary headaches are the main cause of headaches in the community, where the headache itself is the disease [18, 28]. Secondly, brain imaging is not always necessary in the diagnosis of primary headaches, considering the fact that the disease has no impact that leads to macroscopic change in general terms [29]. And finally, primary headache disorders are diagnosed by defining the clinical features of episodes, pain patterns and associated signs and symptoms and then applying them to the established definitions or clinical rules and guidelines for diagnosis, which are formulated by IHS and are accepted worldwide [30].

The majority of *CDSMs* targeting the classification of primary headaches follow the knowledge-based system style, in which knowledge derived from the criteria of IHS and formulated as diagnostic roles by a computer engineer. Researchers usually follow this style when there is an agreed straightforward criterion available worldwide. Meanwhile, there is a lack of online available records of patients with primary headache disorders. This is in turn due to the following: (a) the difficulty of using real-world data because of the privacy policies; (b) the collection of real-world data might be inapplicable, costly or time-consuming; and (c) the real-world data might be unavailable, particularly in the research subjects that newly arose, such as the diagnosis of primary headache disorders using machine-learning methods.

The core concept of the majority of the *CDSMs* was approximately similar regarding the knowledge acquisition, where they have used the criteria of HIS for the classification of headache disorders as a base for deriving the rules. Then, the rules were summarized by a clinical specialist and expressed by a computer engineer. This style is commonly known as a rule-based system, by which the rules are formulated based on a human expert. The basic principle of the rule-based technique is pattern identification followed by a recommendation of what should be done in response, where the rules are a conditional statement that links the supplied conditions to actions or results.

Ideally, the rules are straightforward, understandable and represent the knowledge in near-linguistic form. The rule-based system style can facilitate the separation of knowledge from processing, in addition to allowing incomplete or uncertain knowledge to be expressed and bounded. However, implementing this kind of systems could possibly carry a certain downsides. Firstly, rule-based systems are not able to learn and modify their rules from the experience. Secondly, navigating the categorizations and relationships in a large rule-based system can be complicated and time-consuming. Finally, the most important point is the necessary information needed to derive the rules might consist of more variables than the human mind can accommodate. There are persuasive evidences indicating that the human ability to discover and understand complicated configuration relationships could be limited. Therefore, deriving and formulating the rules with the limited ability of the human mind to manipulate large quantity of information or variables in considering a complex subject such as the criteria of IHS may lead to insufficient rules to deal with the diagnosis of primary headache disorders [2]. Moreover, we would like to pay attention to the fact that the IHS criteria is designed to provide a ground truth for headache specialists, where this classification of headaches provides clear distinct definitions describing many different types of headache. However, these types of headache may share signs and symptoms in real-world scenario, and the types of headache may change over time. This makes the classification of primary headaches not as clear as black or white as shown in the procedural classification function, and there is a grey area in between, which can affect the diagnostic performance, validity and reliability of decisions made by such *CDSMs*.

6.6 Conclusion and Future Plan

In this chapter, we reviewed the literature to explore studies and clinical decision support models that targeted the diagnosis or classification of primary headache disorders. The majority of these studies have followed a knowledge-based system style, in which a computer engineer formulates the diagnostic rules as a set of

IF-THEN-ELSE statements based on clinical guideline or prepared questionnaire. Although the rule-based system style is straightforward and understandable and can represent the knowledge in near-linguistic form, however, it bears many serious downsides, such as the inability to learn and gain knowledge over time, and maintaining categorizations and relationships in a large rule-based system can be complicated. Furthermore, the classification of primary headaches seems to be more complicated and would not successfully achieved using a simple set of rules, as different types of primary headache could share similar signs and symptoms. Currently, we are at the late stage of developing intelligent diagnostic model for primary headaches via completely different approach and employing several machine-learning classifiers to diagnose primary headache disorders using real-world data records of patients with primary headaches. The result of our current study will be available online soon.

References

1. IHS. (2013). The international classification of headache disorders, 3rd edition (beta version). *Cephalalgia, 33*(9), 629–808. https://doi.org/10.1177/0333102413485658.
2. Aljaaf, A. J., et al. (2014). A study of data classification and selection techniques for medical decision support systems. In D.-S. Huang et al. (Eds.), *Intelligent computing methodologies: 10th International Conference, ICIC 2014, Taiyuan, China, August 3–6, 2014. Proceedings* (pp. 135–143). Cham: Springer.
3. Aljaaf, A. J., et al.. (2015). Toward an optimal use of artificial intelligence techniques within a clinical decision support system. In *Proceedimgs 2015 Science and Information Conference (SAI)*, pp. 548–554.
4. Aljaaf, A. J., et al. (2015). A systematic comparison and evaluation of supervised machine learning classifiers using headache dataset. In D.-S. Huang & K. Han (Eds.), *Advanced intelligent computing theories and applications: 11th International Conference, ICIC 2015, Fuzhou, China, August 20–23, 2015. Proceedings, Part III* (pp. 101–108). Cham: Springer.
5. Chen, Y. Y., et al. (2013). Rule based clinical decision support system for hematological disorder. In *Proceeding. 2013 IEEE 4th International Conference on Software Engineering and Service Science*, pp. 43–48.
6. Hasan, R., et al. (2012). An expert system based headache solution. In *Proceedings of 2012 IEEE Symposium on Computer Applications and Industrial Electronics (ISCAIE 2012)*.
7. Al-Hajji, A. A. (2012). Rule-based expert system for diagnosis and symptom of neurological disorders Neurologist Expert System (NES). In *Proceedings of 1st Taibah University International Conference on Computing and Information Technology*, pp. 67–72.
8. Yin, Z., et al. (2013). *A guideline-based decision support system for headache diagnosis* (Series A guideline-based decision support system for headache diagnosis). IOS Press.
9. Dong, Z., et al. (2014). Validation of a guideline-based decision support system for the diagnosis of primary headache disorders based on ICHD-3 beta. *The Journal of Headache and Pain, 15*(1), 40–40. https://doi.org/10.1186/1129-2377-15-40.
10. Yin, Z., et al. (2015). A clinical decision support system for the diagnosis of probable migraine and probable tension-type headache based on case-based reasoning. *The Journal of Headache and Pain, 16*, 29. https://doi.org/10.1186/s10194-015-0512-x.
11. Yin, Z., et al. (2014). A clinical decision support system for primary headache disorder based on hybrid intelligent reasoning. In *Proceedings of 2014 7th International Conference on Biomedical Engineering and Informatics*, pp. 683–687.

12. Simić, S., et al. (2008). Computer-assisted diagnosis of primary headaches. In E. Corchado et al. (Eds.), *Hybrid artificial intelligence systems: Third international workshop, HAIS 2008, Burgos, Spain, September 24–26, 2008. Proceedings* (pp. 314–321). Berlin/Heidelberg: Springer.
13. Simić, S., et al. (2008). Rule-based fuzzy logic system for diagnosing migraine. In *Rule-based fuzzy logic system for diagnosing migraine* (Series: Rule-based fuzzy logic system for diagnosing migraine). Springer-Verlag, pp. 383–388.
14. Eslami, V., et al. (2013). A computerized expert system for diagnosing primary headache based on International Classification of Headache Disorder (ICHD-II). *SpringerPlus, 2*, 199. https://doi.org/10.1186/2193-1801-2-199.
15. Maizels, M., & Wolfe, W. J. (2008). An expert system for headache diagnosis: The Computerized Headache Assessment Tool (CHAT). *Headache: The Journal of Head and Face Pain, 48*(1), 72–78. https://doi.org/10.1111/j.1526-4610.2007.00918.x.
16. Zafar, K., et al. (2013). Clinical decision support system for the diagnosis of migraine and headache. *Journal of Basic and Applied Scientific Research, 3*(10), 119–125.
17. Bigal, M. E., & Lipton, R. B. (2006). Headache classification. In R. B. Lipton & M. E. Bigal (Eds.), *Migraine and other headache disorders, Neurological Disease and Therapy* (1st ed.). Boca Raton: CRC Press.
18. SIGN. (2008). Diagnosis and management of headache in adults: A national clinical guideline. In *Book: Diagnosis and management of headache in adults: A national clinical guideline* (Series Diagnosis and management of headache in adults: A national clinical guideline). Edinburgh: Scottish Intercollegiate Guidelines Network.
19. BASH. (2010). Guidelines for all healthcare professionals in the diagnosis and management of migraine, tension-type, cluster and medication-overuse headache. In Book: Guidelines for all healthcare professionals in the diagnosis and management of migraine, tension-type, cluster and medication-overuse headache (Series: Guidelines for all healthcare professionals in the diagnosis and management of migraine, tension-type, cluster and medication-overuse headache).
20. Friedman, B. W., & Grosberg, B. M. (2009). Diagnosis and management of the primary headache disorders in the emergency department setting. *Emergency Medicine Clinics of North America, 27*(1), 71–87. https://doi.org/10.1016/j.emc.2008.09.005.
21. Katsarava, Z., et al. (2012). Defining the differences between episodic migraine and chronic migraine. *Current Pain and Headache Reports, 16*(1), 86–92. https://doi.org/10.1007/s11916-011-0233-z.
22. Lipton, R. B., et al. (2004). Classification of primary headaches. *Neurology, 63*(3), 427–435.
23. Tepper, S. J., & Tepper, D. E. (2011). Diagnosis of migraine and tension-type headaches. In S. J. Tepper & D. E. Tepper (Eds.), *The cleveland clinic manual of headache therapy* (pp. 3–17). New York: Springer.
24. Stillman, M. J. (2011). Diagnosis of trigeminal autonomic cephalalgias and other primary headache disorders. In S. J. Tepper & D. E. Tepper (Eds.), *The cleveland clinic manual of headache therapy* (pp. 19–36). New York: Springer.
25. IASP. (2012). Trigeminal autonomic cephalalgias: Diagnosis and management. *Book: Trigeminal autonomic cephalalgias: Diagnosis and management* (Series: Trigeminal autonomic cephalalgias: Diagnosis and management).
26. Loder, E., & Rizzoli, P. (2008). Tension-type headache. *BMJ, 336*(7635), 88–92. https://doi.org/10.1136/bmj.39412.705868.AD.
27. Arendt-Nielsen, L. (2015). Headache: muscle tension, trigger points and referred pain. *International Journal of Clinical Practice, 69*(S182), 8–12. https://doi.org/10.1111/ijcp.12651.
28. Morgan, M., et al. (2007). Patient pressure for referral for headache: a qualitative study of GPs' referral behaviour. *The British Journal of General Practice, 57*(534), 29–35.
29. Goadsby, P. J. (2004). To scan or not to scan in headache. *BMJ, 329*(7464), 469.
30. Dodick, D. W. (2003). Clinical clues and clinical rules: Primary vs secondary headache. *Advanced Studies in Medicine, 3*(6C), S550–S555.

Chapter 7
Applications of Educational Data Mining and Learning Analytics Tools in Handling Big Data in Higher Education

Santosh Ray and Mohammed Saeed

7.1 Introduction

"Big data" is a buzzword in today's technological world that everyone is talking about. Each one of us is contributing in generating big data. Big data is generated from heterogeneous data sources such as email, social media, medical instruments, commercial and scientific sensors, financial transactions, satellite and traditional databases, etc. in the form of text, image, audio, video, or any combination of data collected in the form of these. The generation of this huge amount of data is creating an opportunity for organizations to make informed decisions [1]. However, considering the heterogeneous nature and size of big data, management of big data is a not an easy task. Until now, researchers have used data mining to process any large homogeneous datasets. However, to handle the heterogeneity aspects of the big data, the traditional data mining techniques need to be upgraded so that they could handle a different kind of data in parallel. That is why some scholars [2, 3] term "data mining" as "old big data" and "big data" as "new data mining." Big data analytics is being used to examine large datasets containing a variety of data types to uncover hidden patterns, unknown correlations, market trends, customer preferences, and other useful business information. Although big data analytics has been widely used in business environments to predict future trends and consumer behaviors, it has been surprisingly underutilized in the educational environment in general. The six stakeholders in education are learners, educators, educational researchers, course developers, learning institutions, and education administrators. The learners can receive instant and detailed feedback on their interactions with the content they are learning through the learning systems based on big data

S. Ray (✉) · M. Saeed
Khawarizmi International College, Al Ain, UAE
e-mail: santosh.ray@kic.ac.ae; mohammed.saeed@kic.ac.ae

© Springer International Publishing AG, part of Springer Nature 2018
M. M. Alani et al. (eds.), *Applications of Big Data Analytics*,
https://doi.org/10.1007/978-3-319-76472-6_7

analytics. Big data can be used to provide information to students about what they have understood well and what is otherwise. Similarly, practices adopted by high-performance students can be shared with other students so that they can adjust their learning with the system accordingly. Using big data, educators can analyze the overall performance of the class at the macroscopic level, therefore helping them to prepare general strategies for the class. Also, they can analyze performance of an individual student at the microscopic level to find the strengths and weaknesses of that specific student. Accordingly, the educators can focus on the weak points to improve the overall performance of the students. Educational researchers can use a large amount of learner data to propose new learning theories and practices and to test the effectiveness of the proposed theories and models. The course developers can take advantage of the instant availability of a large number of online users and their feedbacks to design new course contents or to modify existing course contents. The learning institutions can use big data to reach potential students for recruiting, or to establish and maintain relations with their alumni. Also, academic administrators can analyze the performance of students from all courses with less effort; they can use these data to measure the effectiveness of new initiatives taken by them to improve the performances of the learners as well as instructors. Accordingly, they can frame policies, implement programs, and adapt the policies and programs to improve teaching, learning, and retention rates. These benefits of big data analytics have generated interests among all the stakeholders in using big data analytics in the learning, administration, and analysis process in the institutions [4].

Analysis of large educational datasets can be done by using the combination of two techniques, namely, educational data mining (EDM) and learning analytics (LA). These techniques develop a capacity for quantitative research in response to the growing need for evidence-based analysis related to education policy and practice [5]. Big data is being used to evaluate the rationality and effectiveness of training programs at universities [6]. Thille et al. [7] studied three different online learning environments: Open Learning Initiative (OLI) at Stanford University and Carnegie Mellon University, Code Webs Project, and massive open online courses (MOOC). They observed that learners and instructors both can benefit from big data. Big data assists instructors in the assessment process by enabling the continuous diagnosis of learners' knowledge and related states and by promoting learning through targeted feedback. Data-enhanced assessment can provide feedback to instructors in designing teaching and assessment strategies in online and offline learning environments. The influence of technology can be seen in many aspects of education from student engagement in learning and content creation to helping teachers provide personalized content and improving student outcomes [8].

The rest of the chapter is organized as follow: Section 7.2 introduces EDM and LA. Section 7.3 briefly describes the major techniques used in EDM and LA. Section 7.4 describes some applications of EDM and LA in education. Section 7.5 describes some tools related to EDM and LA. Section 7.6 presents several case studies of the application of big data in higher education institutions. Section 7.7 provides the conclusion of this book chapter and discusses future directions.

7.2 Educational Data Mining and Learning Analytics

The International Educational Data Mining Society[1] defines EDM as follows: "Educational Data Mining is an emerging discipline, concerned with developing methods for exploring the unique types of data that come from educational settings, and using those methods to better understand students, and the settings which they learn in." Educational data mining applies a combination of techniques such as statistical analysis, machine learning, and data mining to understand learning, administrative, and research issues in the educational sector. Besides these hardcore technologies, EDM requires the knowledge of social network analysis (SNA), psychopedagogy, cognitive psychology, psychometrics, and visual data analytics. In fact, EDM can be drawn as the combination of the three main areas: computer science, education, and statistics [9]. EDM uses techniques and concepts from these different fields in researching, developing, and implementing software tools to identify some patterns in large collections of educational data. Without the aid of EDM techniques, it is impossible for humans to find relevant patterns and data. A review of the developments in the field of educational data mining can be found in [10]. Some of the models used in educational data mining are multiple linear regression model, multilayer perception (MLP) network, radial basis function (RBF) network, and support vector machine [11].

The Society for Learning Analytics Research[2] defines learning analytics as ". . . the measurement, collection, analysis and reporting of data about learners and their contexts, for purposes of understanding and optimizing learning and the environments in which it occurs." The objective of LA is to analyze the large dataset and provide feedbacks that have an impact directly on the students and the instructors and the details of the learning process. For example, information based on analyzing the students activities and interaction among students can be used to provide a personal recommendation to students' learning or change in the course material [12–15].

Although LA and EDM share many attributes and have some similar goals and interests, there are some major differences between them as described in [16]. LA has an origin in Semantic Web, intelligent curriculum, and systemic interventions, while EDM has origin in educational software, student modeling, and predicting course outcomes. Researchers in LA use statistics, visualization, social network analysis, sentiment analysis, influence analytics, discourse analysis, concept analysis, and sense-making models more frequently, while researchers in EDM rely more on classification, clustering, Bayesian modeling, relationship mining, and discovery with models. LA focuses on the description of data and results, while EDM focuses more on procedures and techniques.

---

[1] http://www.educationaldatamining.org/

[2] http://www.solaresearch.org/

7.3 Methods in Educational Data Mining and Learning Analytics

Some of the most commonly used methods in EDM and LA include:

Classification and Prediction Classification and prediction refer to a group of data mining methods that search and identify a relationship between independent variable (target variable) and dependent variables based on past values of data. Classification methods define a fixed number of classes and predict objects to fall in one of the defined classes on the basis of a trained model. Prediction method usually tries to predict the missing value of the independent variable using continuous function models. Classification and prediction methods have been used in education for forecasting student performance on the basis of their behaviors in an online environment [17, 18].

Clustering Clustering refers to identifying data points that are similar in some respect so that a full dataset can be split into various categories of small datasets. In a typical clustering process, certain kind of distance measure is used to decide how similar data points are. In the educational field, clustering can be used for grouping similar course materials or grouping students based on their learning and interaction patterns and grouping users for purposes of recommending actions and resources to similar users [19, 20]. The varied and voluminous nature of online learning environments provides good opportunity to use clustering techniques in analyzing online data. Clustering can be used in any domain that involves classifying, even to determine how much collaboration users exhibit based on postings in discussion forums [21].

Outlier Detection The outlier detection methods refer to the process of identifying data points that are significantly different than the rest of data. The volume of data points detected as outlier is usually much larger or smaller than other data. Outlier detection method can be used to identify abnormal fall in the performance of the students or instructors, to identify students at the extreme ends of the performance spectrum [22].

Relationship Mining Relationship mining refers to the process of discovering relationships between variables in a dataset and encoding them as rules for later use. There are several types of relationship mining including the most popular association rule mining. In the educational field, relationship mining can be used to identify relationships between students' poor performance and their behavior during the learning process. This relationship, then, can be used to build recommendations for content that is likely to be interesting, or for making changes to teaching approaches [23].

Social Network Analysis The social network analysis considers individuals as nodes of a graph and relationship between individuals as edges between nodes. The aim of the social network analysis is to determine and understand the relationships between entities in a networked environment such as discussion forums or social media. In education, SNA can be used to detect and understand the user interaction with a communication tool. It can also be used to determine the contribution of each member in collaborative projects. Social media analysis can provide information about the centrality of the nodes, i.e., which student(s) in the network played the more vital role in connecting with other students [24].

Process Mining The process mining provides a visual representation of useful knowledge from event logs of an information system. Therefore, the process mining can be used to mine learning management systems for visual presentation of student performance in various assessments during a course [25].

Text Mining The text mining refers to set of processes used in analyzing unstructured texts and deriving high-quality information from raw text. Some of the applications of text mining include text categorization, text clustering, named-entity extraction, production of granular taxonomies, sentiment analysis, document summarization, and entity relation modeling. The text mining methods have been used to analyze the contents of discussion boards, forums, chats, web pages, documents, and so forth [26].

Distillation of Data for Human Judgment Distillation of human judgment refers to the process of representing data using visualization and interactive interfaces to enable a human to quickly identify or classify features of the data. Distillation of data for human judgment can help humans to easily identify features of student learning actions, or patterns in student behaviors by analyzing a large amount of educational data at once [27].

Discovery with Models Discovery with models is a technique that uses a previously validated model of a phenomenon (using prediction, clustering, or manual knowledge engineering) as a component in another analysis such as prediction or relationship mining [28]. In education, discovery with models can be used to discover relationships between student behaviors and students' characteristics or contextual variables, the analysis of research questions across a wide variety of contexts, and the integration of psychometric modeling frameworks into machine-learning models [29].

7.4 Applications of EDM and LA in Education

Learning analytics and educational data mining match students background information of their interaction with the learning management systems. It tries to comprehend how students interact with university resources, their learning styles, likely performance, and perhaps most pertinently, how likely they are to complete their studies successfully [30, 31].

Papamitsiou and Economides [32] did a study on the use of learning analytics and data mining in education. They listed the basic objectives of the major research done between 2008 and 2013 in the field of educational data mining. These objectives were student/student behavior modeling, prediction of performance, increase (self-) reflection and (self-) awareness, prediction of dropout and retention, improve feedback and assessment services, and recommendation of resources. Based on these objectives, some tasks where EDM and LA techniques have been applied are described below.

Predicting Student Performance Governments across the world are reducing the funding of higher education institutions. In order to fill the gap between decreasing revenue and increasing expenses, the institutions are under tremendous pressure to increase the new enrollments and retain the existing students. One of the several measures adopted by the institutions is to predict the possible dropout students at the early stages so that the necessary remedial measures can be preplanned. Big data can help in predicting possible students' dropout by identifying and analyzing various parameters such as cumulative grades in prerequisite courses; marks obtained in previous quizzes, tests, and assignments; students' participation in activities; and multimodal skills. Once the set of possible dropout students are identified, the student retention rate of the institute can be improved by taking proactive measures such as one-to-one tutoring, the arrangement of remedial classes, etc. [11, 33]. In order to predict students' dropout at early stages in e-learning courses, Lykourentzou et al. [34] applied a combination of three machine learning techniques, namely, feed-forward neural networks, support vector machines, and probabilistic ensemble simplified fuzzy ARTMAP on detailed students' logs from a learning management system. Dekker et al. (2009) [35] used various classification algorithms to predict students' dropout and identify factors of success based on the first-year enrollment data itself.

In addition to detection of possible students' dropout, researchers used big data techniques to measure the satisfaction, motivation, and performance of the students by analyzing their activities during lectures and exams. Kizilcec et al. [36] classified learners according to their interactions with course content (video lectures and assessment) in learning activities in massive open online courses. They analyzed behaviors engagement patterns of students in three different computer science courses and compared clusters based on learners' characteristics and behavior. Giesbers et al. [37] investigated the relationship between students' interaction with synchronous tools in online education and performance on a final exam. Another objective of this research was to determine whether actual usage of synchronous tools increases the motivation to participate in online courses that support these tools. Abdous et al. [33] and He [38] examined the relation between interactions within a live video streaming (LVS) environment and students' final grades in order to predict their performance, discover behavior patterns in LVSs that help increase performance, and understand the ways students are engaged into online activities. Dejaeger et al. [39] studied how retention of students can be increased by enhancing students' satisfaction. The authors investigated students' satisfaction using several class and training variables such as the perceived usefulness of training, perceived training efficiency, etc.

Realizing the limitations of statistical models in the accurate prediction of student performances, Xing et al. [40] used learning analytics, educational data mining, and human-computer interaction (HCI) theory to develop a model to predict the final performance of the student. They used data of 122 students from an online Math course to generate categories of student performance. Zacharis et al. [41] used a multivariate model to predict the "at-risk" students. They studied 29 variables used in a Moodle-based learning environment and found only 14 variables to be significant. However, the results suggested that only four variables – reading and posting messages, content creation contribution, quiz efforts, and number of files viewed – provide an 81.3% accuracy in prediction. Niemi and Gitin [42] used some nonacademic parameters (e.g., age, gender, race, marital status, military status, previous college education, estimated family financial contribution, and the number of transfer credits from another university) besides the academic variables (final exam score, discussion participation, project scores, and other assignment scores) to predict the possible rate of students' dropout. They studied and applied learning analytics to a database of 14,791 students enrolled in a fully online program. They observed that married people or people working with the military have a lower probability to drop the course. The use of textbooks and e-books plays a very crucial role in improving the performance of the students. Junco and Clem [43] used linear regression analysis technique to link the usage of digital textbooks with the performance of students. They stressed that this easily traceable variable is a much stronger indicator of performance than other variables including the past performance of the student. Mouri et al. [44] also used Bayesian network to analyze approximately 330,000 logs from 99 first-year students to establish the relationship between the usage of e-books and performance of students.

Educating Students Using Big Data The twenty-first century has witnessed the integration of ICT into teaching and learning. The educators are using a number of online and offline tools to create quality and easily understandable content to learners. One of the most adopted ICT tools in higher education institutions during the first decade of this century was web-based learning management systems (LMS). LMSs such as Blackboard and Moodle are helping educators in bringing together learning contents and resources besides other administrative jobs such as assessment of student's work, etc. LMSs are proving to have limitations in their monitoring capabilities. Therefore, the second decade of this century is witnessing the emergence of distributed heterogeneous tools used by all the stakeholders of the learning process [12]. These systems are embedding data mining techniques to collect the required data, analyze them, and suggest the appropriate actions. The big data can help in tracking the time taken by the students to learn a particular concept. This will be an indicator of the level of difficulty of the concept provided in the study material, or it can help to determine the learning ability of the students. For example, researcher Paulo Blikstein [45] examined a sample of college students in a computer programming class to see how they solve a modeling assignment. He used NetLogo software to maintain logs of all user actions from button clicks and keystrokes to code changes, error messages, and use of different variables. He found that error rates progressed in an "inverse parabolic shape" as students tried things and made a lot of mistakes initially, and then progressed through problem-solving until they had developed the correct model. Researchers are also investigating the techniques to analyze the students' moods instead of simply analyzing the computer commands used by the students. This can help to assess students' interest in the course generally more deeply. Moridis and Economides [46] proposed a formula-based method and a neural network-based method to automatically collect the affect state of the student during learning.

A typical large size class, especially in e-learning environments, consists of students from different knowledge backgrounds whose educational requirements are quite different. Offering the same learning path of content to all of them can negatively affect their overall performance. It has been observed that the personalized learning can make learning activities more effective by suiting the learning process to learners' needs and enhancing learner motivation [47, 48]. Educational data mining techniques can be used to create a customized learning environment in which students can be provided personalized learning paths for optimizing their performance [49]. Some data mining techniques such as clustering, associate rules [50], and feature selection [51] have been applied in developing personalized learning systems and increasing individual learning performance. Due to their inherent strength such as displaying results in user-understandable formats, ability to analyze both continuous and discontinuous variables efficiently, and flexibility with type and scale of databases, decision trees have been popular in designing personalized learning contents [52, 53]. Designing the personalized learning content path requires accurate estimation of the learning abilities of the students at various stages. Researchers have considered this issue also and used statistical techniques such as Gaussian approximation method to estimate the learning ability of the students in a typical web-based learning environment [54].

Assessment of Students' Learning There are several issues with the traditional way of evaluating students' learning. However, the use of EDM in the assessment of learning can result in faster progress as EDM can provide a real-time and continuous assessment [55]. Instead of conducting a periodic exam with a fixed set of questions for all students, big data can be used to create dynamic test according to the knowledge of the student. This can enable the instructor to find out the precise weakness of each student, and the instructor can prepare study plan tailored to the needs of the individual students. Romero et al. [56] used association rules mining to improve quizzes and courses. They used different objective and subjective rule evaluation measures to select the most interesting and useful rules. Based on the selected rules, the proposed system provides feedback to the instructors to improve quizzes and courses. Data mining methods such as clustering, classification, and association analysis have been used to study how well the questions in the test and the corresponding elaborated feedback were designed or tailored towards the individual needs of the students [57].

Teaching and Research Big data techniques can be useful to identify the academic resources to increase the awareness of the instructors. Researchers can leverage the EDM and LA techniques to explore and research large data made available by MOOC. The analysis of textual and video data can provide many insights for instructors. For example, after analyzing the video data and performance of the students, researchers discovered that the presentation of the instructor's face in video lectures influences attrition and achievement rates and they found heterogeneous effects on attrition [58]. An in-depth analysis of demography of students enrolled in MOOCs can provide researchers with heterogeneous samples of people from traditionally underrepresented demographic and sociocultural groups in more narrowly obtained educational datasets. The researchers can leverage this data to conduct large-scale field experiments and evaluate multiple theories at minimal cost [59]. Racial discrimination is a big issue in educational establishments. Big data can help in identifying racial discrimination. To achieve this objective, Baker et al. [60] planted messages in discussion forums across 126 MOOCs (1008 messages in total, eight per course) and randomly assigned learner names to be evocative of different races and genders. They found evidence of discrimination in the behavior of instructors and students. For example, instructors wrote more replies for white male names than for white female, Indian, and Chinese names. Peer pressure is another critical issue in any educational environment. Kizilcec et al. [61] conducted a research to study how the level of transparency about the peer grading process affects learners' trust in peer grading. They concluded that fair and transparent peer grading procedure can promote resilience in trust of learners who received a lower than expected grade. However, the downside of peer pressure was studied by Rogers and Feller [62] who found that exposure to exemplary peer performance causes attrition, due to the upward social comparison that undermines motivation and expected success.

7.5 EDM and LA Tools

There are a number of free and commercial data mining tools available today. One such list of data mining tools can be found at SourceForge.[3] However, not all of these data mining tools are designed to meet the requirements of educational data mining. Some of the tools used in educational data mining are described in this section.

Education Prediction Rules (EPRules) EPRules [63] is a java-based graphical tool used to solve the prediction rule discovery in adaptive systems in a web-based learning environment. This tool can be used even by course developers or teachers who are not expert in data mining. The data input component of this tool

---

[3]https://sourceforge.net/projects/gait-cad/files/wiley_irdmkd_data_mining_tools/tools.xls/
download

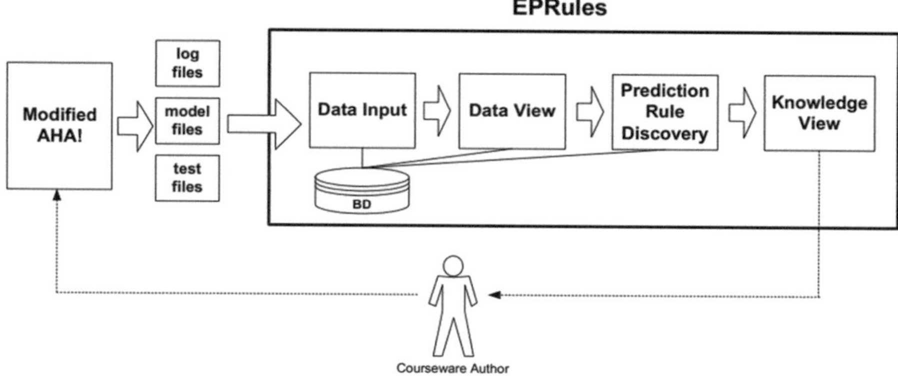

Fig. 7.1 EPRules [64]

allows to open an existing database or create a new one using course usage file. The data view component helps in visualizing the students' course usage data. The prediction rule discovery component (shown in Fig. 7.1) allows selecting one of the several rule discovery algorithms, to choose the specific execution parameters for the chosen algorithm, to select the subjective restriction (such as the number of chapters or number of students), and to choose the objective evaluation function. The last component knowledge view displays the discovered rules, conditions of the rules, and evaluation parameters.

Graphical Interactive Student Monitoring (GISMO) GISMO[4] is a graphical interactive monitoring tool that provides useful visualizations of students' activities in online courses to instructors. With GISMO, instructors can examine various aspects of distance students, such as attendance to courses, reading of materials, and submission of assignments. GISMO, in tandem with Learning Management System Moodle, can provide comprehensive visualizations that give an overview of the whole class, not only a specific student or a particular resource [27] (Fig. 7.2).

Tool for Advanced Data Analysis in Education (TADA-Ed) It is a data mining platform that helps teachers to mine and visualize students online exercise work such as students' interactions and answers, mistakes, teachers' comments, and so on [65]. TADA-Ed contains preprocessing facilities so that users can transform the existing database tables to a format that, when used with a particular data mining algorithm, can generate meaningful results for the teacher (Fig. 7.3).

Synergo/Collaborative Analysis Tool (ColAT) Synergo [67, 68] is a synchronous collaboration-support environment that allows a group of students to chat and share problem-solving activities. It keeps track of user operations and allows to analyze

---

[4]http://gismo.sourceforge.net/index.html

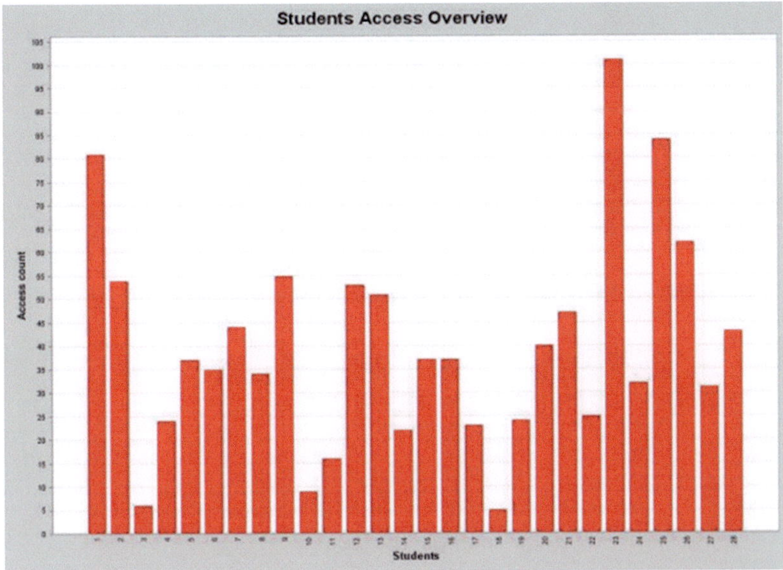

Fig. 7.2 A graph by GISMO reporting the student's accesses to the course

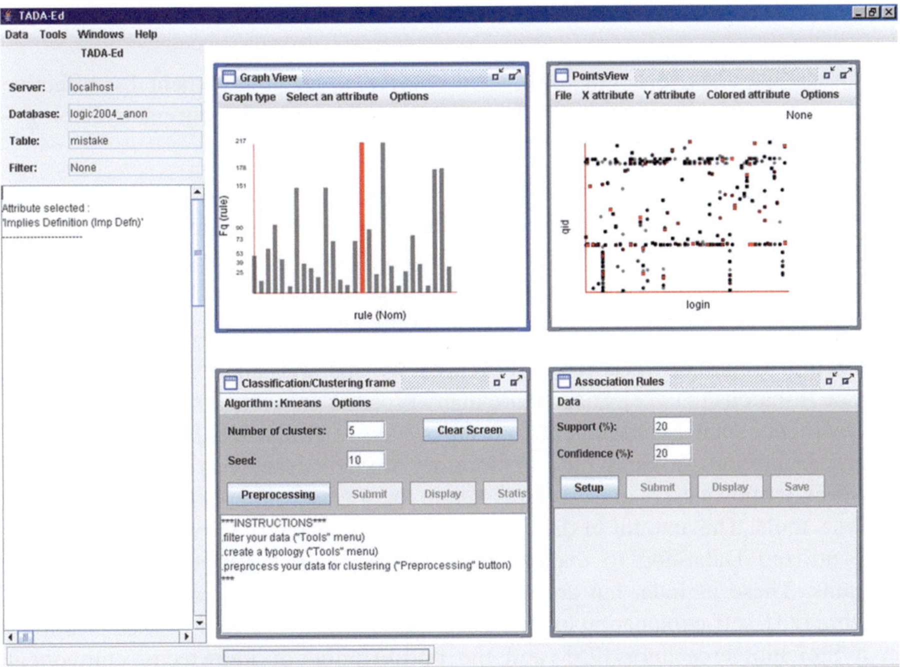

Fig. 7.3 A general overview of TADA-Ed [66]

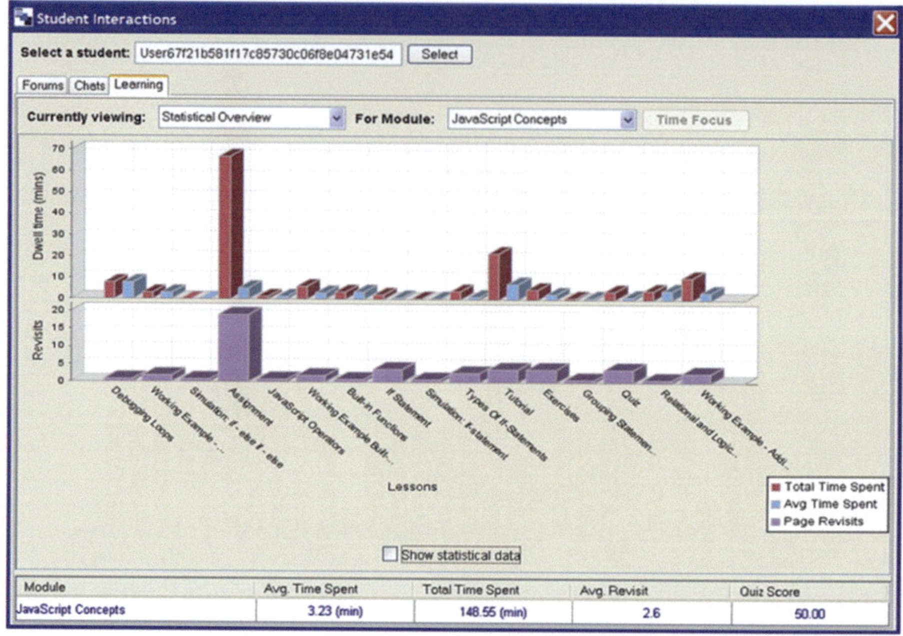

Module	Avg. Time Spent	Total Time Spent	Avg. Revisit	Quiz Score
JavaScript Concepts	3.23 (min)	148.55 (min)	2.6	50.00

Fig. 7.4 LOCO-Analyst [70]

the usage log files in a graphical form. ColAT is an environment for effective analysis of interrelated multiple data that may be collected during technology-supported learning activities.

LOCO-Analyst LOCO-Analyst is an educational tool that provides teachers with feedback on the relevant aspects of the learning process taking place in a web-based learning environment. It provides feedback on student activities during the learning process, usage and comprehensibility of the learning content provided by the teacher, and contextualized social interaction among students [69] (Fig. 7.4).

DataShop DataShop[5] is a free data repository and web application for learning science researchs [71, 72]. It can store many types of data from interactive learning environments such as intelligent tutoring systems, virtual labs, simulations, and games. DataShop provides only exploratory statistical analysis of learning data. However, it allows users to export the data in the formats suitable to other statistical analysis tools. The amount of data in DataShop is constantly growing. Researchers have utilized DataShop to explore learning issues in a variety of educational domains. These include, but are not limited to, collaborative problem-solving in Algebra [73], self-explanation in Physics [74], the effectiveness of worked examples in a Stoichiometry tutor [75], and the optimization of knowledge component learning in Chinese [76].

---

[5]https://pslcdatashop.web.cmu.edu/about/

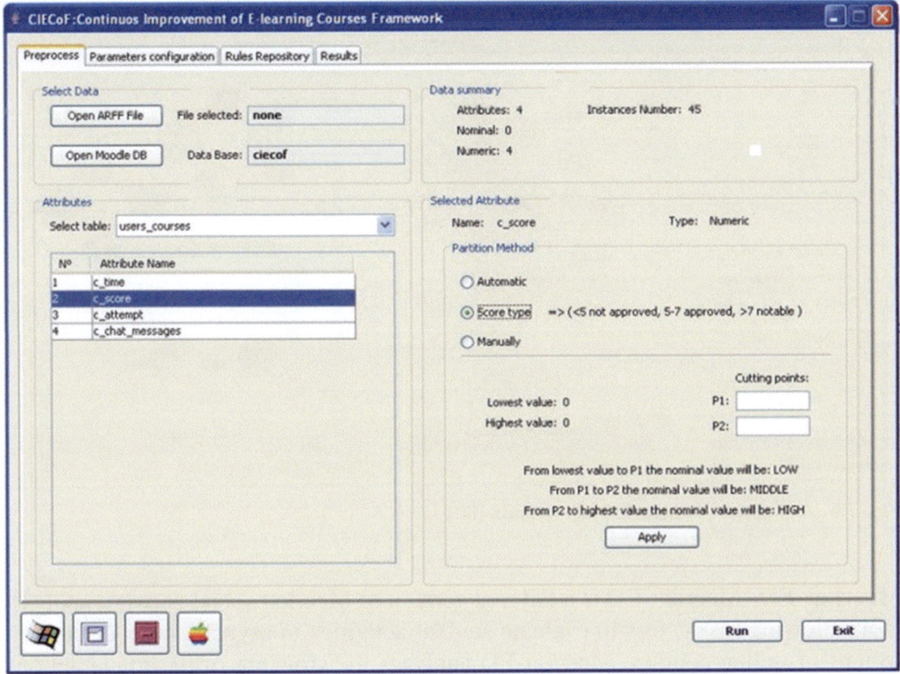

Fig. 7.5 CIECoF [77]

Continuous Improvement of E-Learning Courses Framework (CIECoF)
CIECoF is a tool intended to help instructors to discover, score, and share information with other instructors teaching similar courses [77]. This tool is based on client-server architecture. The client subsystem analyzes student usage data in the learning system using associate rules. The server subsystem enables to score and share the discovered rules by other teachers of the similar course (Fig. 7.5).

Student Activity Monitoring Using Overview Spreadsheets (SAMOS) SAMOS is an information system that facilitates the automatic generation of weekly monitoring reports derived from data contained in online collaborative learning environments [78]. It allows instructors to view classification of students according to the activity level which can be helpful in identifying at-risk students. It uses Excel's numerical, graphical, and programming capabilities to generate weekly reports from the student activities stored on the server of the learning management system and sends them to the instructors through email.

PDinamet This is a web-based adaptive learning system that consists of several types of learning resources. Each resource is presented by a set of characteristics such as difficulty level and learning objectives [79]. PDinamet contains personal and academic information (such as performance in the previous test) of students and recommends learning resources for students.

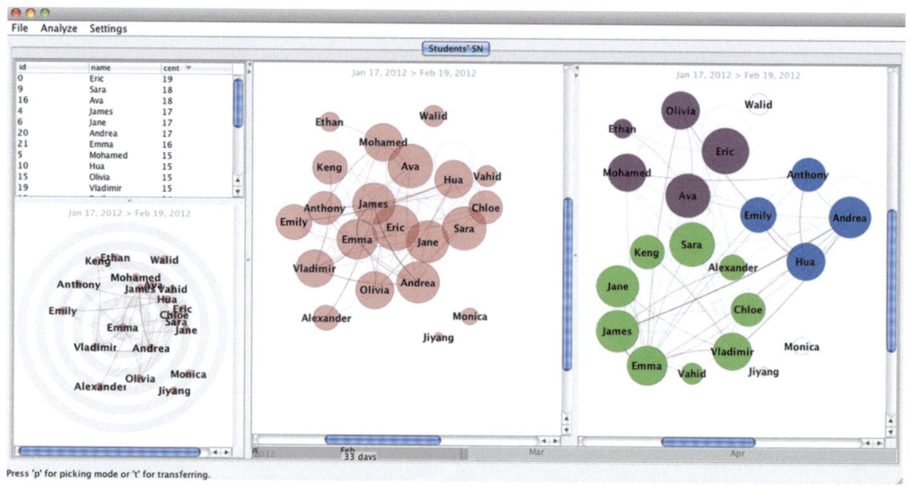

Fig. 7.6 A snapshot of Meerkat-ED toolbox [80]

Meerkat-ED Meerkat-ED is a tailored version of Meerkat social network analysis tool[6] allowing instructors to evaluate student activities in asynchronous discussion forums of online courses. Meerkat-ED analyzes the structure of these interactions using social network analysis techniques including community mining. It prepares and visualizes overall snapshots of participants in the discussion forums, their interactions, and the leader/peripheral students in these discussions. Moreover, it analyzes the contents of the exchanged messages in these discussions by building an information network of terms and using community mining techniques to identify the topics discussed. Meerkat-ED creates a hierarchical summarization of these discussed topics in the forums, which gives the instructor a quick view of what is under discussion in these forums. It further illustrates how much each student has participated in these topics, by showing their centrality in the discussions on that topic, the number of posts, replies, and the portion of terms used by that student in the discussions [24] (Fig. 7.6).

Knowledge Building Discourse Explorer (KBDeX) KBDeX[7] is a discourse analysis tool based on the relation between the words and discourse unit [81]. It helps collaborative learning researchers to visualize network structures of discourse based on a bipartite graph of words vs discourse units. This can be used to compare coefficients across different phases of collaborative learning between groups. KBDeX supports stepwise analysis to calculate each individual's contribution (Fig. 7.7).

---

[6]http://www.amii.ca/meerkat/

[7]http://www.kbdex.net/

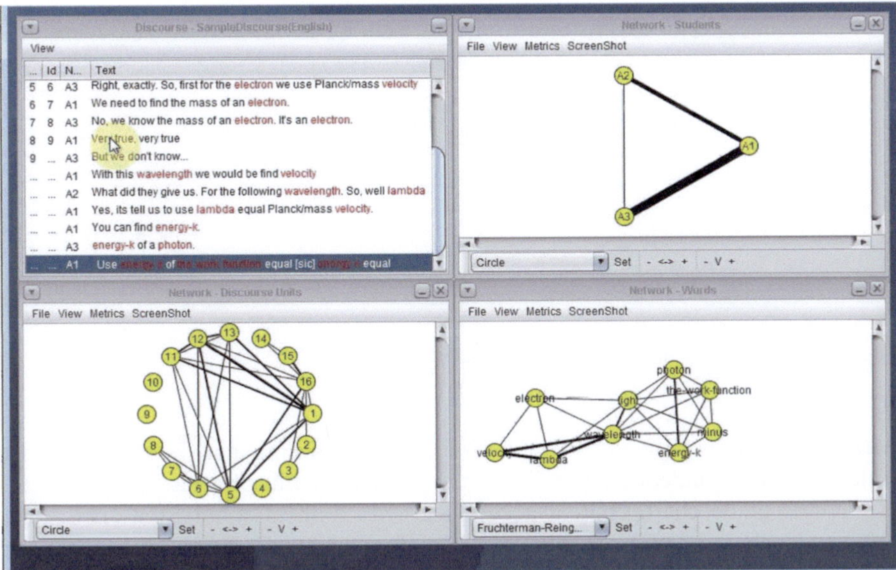

Fig. 7.7 The main view of KBDeX [82]

Moodle Data Mining (MDM) Tool MDM[8] is a free available learning analytics tool developed using PHP and can be easily integrated into Moodle as a module for a specific course. It supports several tasks such as selection, data preprocessing, and data mining from Moodle courses [83]. This tool can be useful in providing instructors with feedback about how students learn within Moodle courses. Data preprocessing component of the tool allows the instructor to load raw excel data, edit, anonymize, discretize, and split the data. Data selection component enables specific data (summary, logs, forum discussions, grades, etc.) chosen by the instructor from the Moodle course. The data mining component runs knowledge discovery algorithms for clustering and classification of data (Fig. 7.8).

Academic Analytics Tool (AAT) AAT [84] assesses and analyzes students' behavior data in learning systems. Modern learning management systems store vast amounts of data about every action students take while interacting with their courses, instructors, and learning materials/activities. However, a regular user does not know how to access these behavioral data from the learning management system. AAT is an interface based tool that allows users to ask questions related to user behavior or study materials in natural language. AAT generates graphical representations of the answers of the user questions that can be easily understood and used by regular users such as course instructors.

---

[8]http://www.uco.es/grupos/kdis/index.php?option=com_content&view=article&id=23&Itemid=60&lang=en

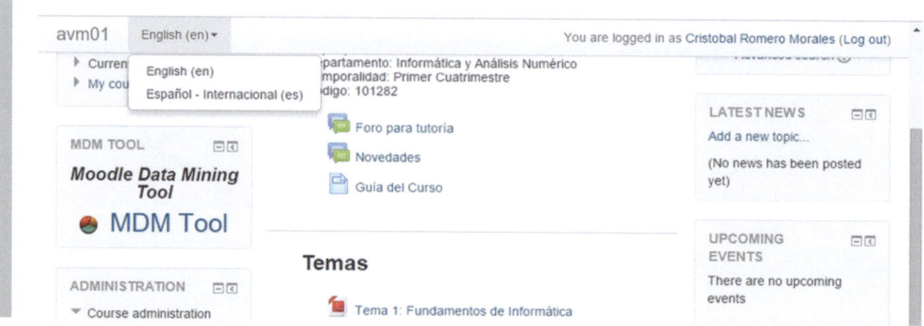

Fig. 7.8 MDM tool block in an example course [83]

Analytics Graphs Analytics Graphs is a Moodle learning analytics plug-in[9] that collects existing student activity data from Moodle and displays it in visual form. It supports grade chart, content access chart, assignment submission chart, quiz submission chart, and hits distribution chart. By using these charts, instructors can easily notice the things which could have gone unnoticed otherwise.

CVLA CVLA [85] is a Moodle-based tool that integrates analytics techniques to produce a custom Moodle report. This system uses multiple datasets and analytics techniques in a single interface for presenting data to learners and educators. It is integrated into Moodle as a module and provides social network analysis and classification algorithms for predicting assignment submission.

E-Learning Web Miner E-learning Web Miner is a data mining tool developed at the University of Cantabria [86]. It helps instructors to discover and analyze students' behavior in distance learning programs by analyzing navigational and demographic data. It reveals students' behavior profiles and models how they work on virtual courses so that instructors can use these data to improve their courses. It is a web-service that provides visualization graphs, clustering, and association algorithms.

IntelliBoard.net IntelliBoard.net[10] extracts the statistical data gathered and available in Moodle and Totara, and presents it on a dashboard in the form of printable charts, graphs, and multiple-format reports. IntelliBoard.net provides multiple reports, analytics, and notifications that keep learners focused, and provide data that can improve learning methodologies. By providing the learner's status summary such as success and failure rates, assignment due status, etc., the IntelliBoard.net can help institutions in identifying at-risk students (Fig. 7.9).

---

[9]https://moodle.org/plugins/block_analytics_graphs

[10]http://www.intelliboard.net/

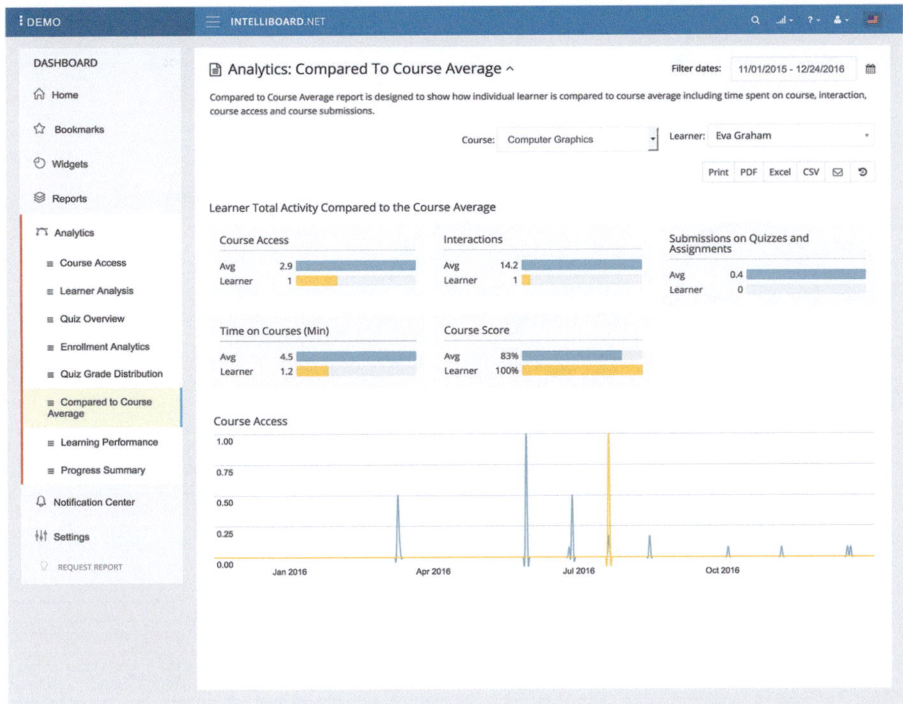

Fig. 7.9   IntelliBoard.net

MOClog   MOClog [87] is a monitoring system that helps to analyze log files of the Moodle LMS more effectively and efficiently. For example, log file analyses can help in better understanding whether the courses provide a sound learning environment (availability and use of discussion forums, etc.) and implement best practices in online learning (students provide timely responses, teachers are visible and active, etc.). It allows analyzing the use of the contents in the online courses from a didactical point of view, thus going deeper than simply counting and visualizing the numbers of posts and clicks.

SmartKlass   SmartKlass[11] is a learning analytics plug-in that can be used by any virtual learning system to measure and analyze the learning process at any time. Its objective is to empower teachers to manage the evolution of the students in an online course. It is an open-source and multi-platform learning analytics dashboard plug-in. It allows teachers to see a global view of the performance of the students, check the evolution of any course, and control and check an alarm system to send messages to the students. Similarly, it enables students to view their performances, see the evolution of the course, and receive or send alert messages (Fig. 7.10).

---

[11] http://klassdata.com/smartklass-learning-analytics-plugin/

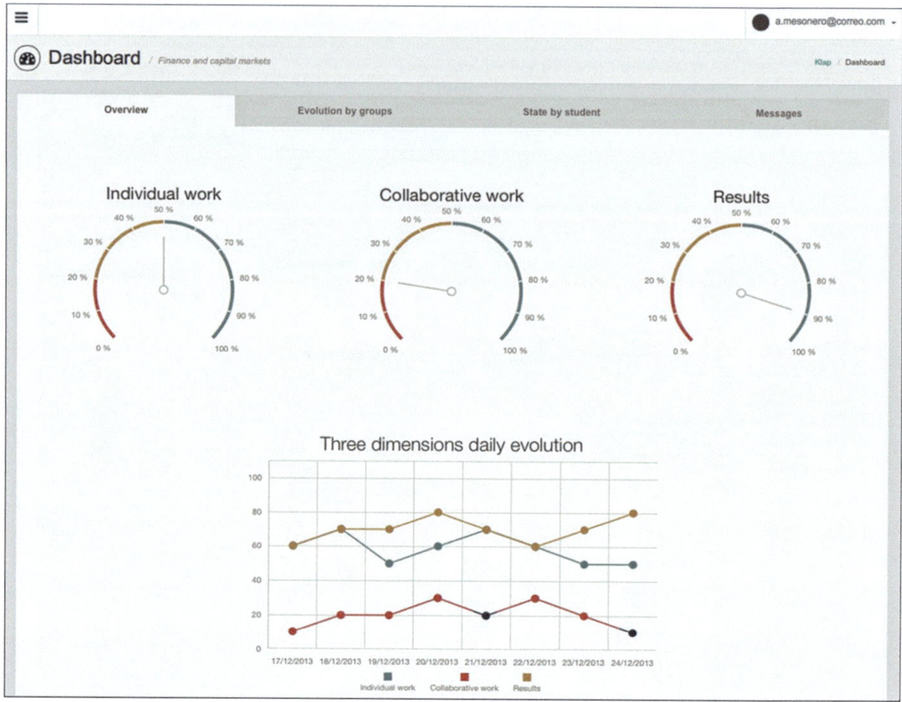

Fig. 7.10 SmartKlass

Social Networks Adapting Pedagogical Practice (SNAPP) SNAPP [88] is a bookmarklet that visualizes the evolution of participant relationships within discussion forums. It displays timeline and number of the posts, chart of all users, their posts, and how other users have replied to those posts among other things.

7.6   Case Studies

In this section, several case studies of the application of big data in higher education institutions are discussed and evaluated.

- The Open University (OU)[12] started a pilot project in 2014 to provide an early prediction of "at-risk" students and overall class engagement. They constructed four predictive models (Bayesian classifier, classification and regression tree (CART), k-nearest neighbors (k-NN) with demographic/static data, and k-NN with virtual learning environment data) based on machine learning. Initially

---

[12]https://en.wikipedia.org/wiki/Open_University

started with two courses, this project was later extended to other 10 courses at different levels. After successful implementation and testing of this project, the lecturers had the necessary information so that they could focus their efforts on struggling students and amend course material proven ineffective earlier [89]. Learning analytics gives them the opportunity to do so in real time, without the delay usually associated with student feedback and outcomes. A preliminary evaluation of this pilot project has shown retention rates increased by 2.1% on average compared to the previous year. Not only will this improved university finances – the higher retention rates generated an estimated £1.8 million in additional income for the OU [90].

- In 2014, a learning analytics initiative was rolled out at Nottingham Trent University, UK [91]. In this initiative a student dashboard designed using learning analytics methods was rolled out throughout the institute. The dashboard calculates student engagement score from virtual learning environment access, library usage, card swipes, and assignment submissions. The dashboard was initially tested with 4 courses, 40 tutors, and over 500 first-year students. After the success of the pilot project, the project was implemented throughout the University. Tutors are prompted to contact students when their engagement drops off. The University found that 27% of students with access to their own dashboard changed their behavior – for example, by increasing their attendance – while one third of tutors contacted students as a direct result.

- Odette School of Business, University of Windsor, Canada, has developed the Assurance of Learning (AOL) analyzer [92] to assess the outcomes of their undergraduate and postgraduate programs in business and management. They are using R programming environment to implement the AOL analyzer. The analyzer has two subsystems: document management system and data management system. The document management system assists in managing various documents related to quality and education. The data management system processes the data related to these documents. In addition to informing faculty on the actual percentage of students who obtained 70% or more on the AOL test, the AOL Analyzer also informs faculty about the score achieved by 70% or more students. Should students tested for a learning outcome fail to achieve the threshold, the AOL Analyzer informs faculty what percentage achievement level 70% of students actually achieved. This information supports discussion on the identification and implementation of program improvements to assure learning achievement improvement.

- Plan Ceibal (www.ceibal.edu.uy, http://www.fundacionceibal.edu.uy/en) is a national policy program in Uruguay under which each student and teacher is provided with a laptop and internet. Plan Ceibal currently offers a set of educational software platforms for teaching, learning, training, hosting, exchanging, and creating information. Virtual learning environments (VLE) at Plan Ceibal allow real-time interaction between students and their teachers and peers through a variety of resources and exercises, discussions, or instant messaging. These VLEs generate massive amounts of data on the progress and style of students' learning. In order to analyze the massive data, a big data center for learning analytics is being planned [93]. Applications are being developed to exploit the learning analytics and big data to support the education system. One of such system is 3600 user profile [94] that is intended to build a comprehensive user online profile. Though the system is in design stage, it aims to use advanced EDM and learning analytics techniques to support the work of educators while they plan their teaching and to provide relevant data for the learners regarding their performance.

- Kim et al. [95] conducted a study between September 2015 and December 2015 onto a set of over 650 students in Pusan University, Korea. They used learning analytics techniques to analyze the data from an online Naver Cafe to address the relationship between the number of logins and grades for each individual student, to understand challenging problems that most students consider difficult by analyzing the page hits, and to understand learning styles of individual students by monitoring and analyzing the number of Café members, site hits per day, and paths leading to the Café. The analysis showed a strong correlation between the involvement in the Café activity and the grade. They also observed that the problems and subjects that most of the students consider challenging could be identified by analyzing the hits of the popular web pages. Also, the learning style of individual students could be determined based on the changes in the number of members over time, postings, page views, and paths leading to the Cafe.

7.7 Conclusion and Future Directions

In this chapter, we discussed techniques and applications of education data mining and learning analytics. We described some common tools which can be useful to researchers, instructors, administrators, and eventually students through analyzing the behaviors and performance of the students. We also presented several case studies of applications of learning analytics and data mining techniques in various educational institutions across the globe.

The existing learning management systems and supporting tools are supporting educational institutions in analyzing performances of the students. But, we cannot say the same about the analysis of satisfaction of the students with course curriculum, faculty performance, and LMS tools. Most of the educational institutions have already established feedback system (e.g., student survey) to capture opinions and views of the students. But usually results from these feedback systems are biased. Many of the students are either afraid to be honest or don't feel the urge to express their opinions. In order to solve this issue, feedback should be taken not only in a completely anonymous manner but also at the flexible timing. Students should be given chance to express their opinions at the time and place of their choice [96]. Social media provides an excellent platform to express opinions in anonymous and flexible manner. Users may freely express their feeling about education systems, course curriculum, faculties, and learning management systems. Analysis of these almost true feedbacks will give insight into what students really think about our education systems. Researchers need to focus on developing tools to analyze these huge feedbacks scattered over social media to gain more insights about the performance of specific lectures and professors and the usefulness of learning management systems and institutions. Some social media analysis tools, for example, sentiment viz.[13], are available, but they have not been designed to meet the requirements of educational institutions.

Some recent government-supported researches are underlining the importance of big data in higher education and research. An OECD (2013) report suggested that big data may be the foundation on which higher education can reinvent its business model and bring together the evidence to help make decisions about educational outcomes [97]. Based on these types of researches, governments are planning future steps to improve education by using big data. Recently conducted workshop on data-intensive research in education [98] suggested the following steps to improve educational levels: (1) mobilize communities around opportunities based on new forms of evidence; (2) infuse evidence-based decision-making throughout a system; (3) develop new forms of educational assessment; (4) reconceptualize data generation, collection, storage, and representation processes; (5) develop new types of analytic methods; (6) build human capacity to do data science and to use its products; and (7) develop advances in privacy, security, and ethics. These initiatives by the government-funded agencies will accelerate the much-needed reform process in the education sector.

References

1. Erevelles, S., Fukawa, N., & Swayne, L. (2016). Big data consumer analytics and the transformation of marketing. *Journal of Business Research, 69*, 897–904. https://doi.org/10.1016/j.jbusres.2015.07.001.
2. Giacalone, M., & Scippacercola, S. (2016). Big data: issues and an overview: In some strategic sectors. *Journal of Applied Quantitative Methods, 11*(3), 1–17.
3. Zhou, R. R. (2016). Education web information retrieval and classification with big data analysis. *Creative Education, 7*, 2868–2875. https://doi.org/10.4236/ce.2016.718265.
4. Dawson, S., Gasevic, D., Siemens, G., & Joksimovic, S. (2014). Current state and future trends: a citation network analysis of the learning analytics field. In *Proceedings of the Fourth International Conference on Learning Analytics & Knowledge* (pp. 231–240). New York, USA: ACM New York.
5. Besbes, R., & Besbes, S. (2016). *Cognitive dashboard for teachers professional development*. Qatar Foundation Annual Research Conference Proceedings 2016: ICTPP2984 https://doi.org/10.5339/qfarc.2016. ICTPP2984.
6. Yanfeng, Y. U. E., & Da, L. I. U. (2016). Evaluation of Different Training Programs of Innovative Education in Top International Universities using Big Data Analysis. *International Journal of Simulation—Systems, Science & Technology, 17*(42), 1, 5P–5.
7. Thille, C., Schneider, E., Kizilcec, R., Piech, C., Halawa, S., & Greene, D.K. (2014). The future of data-enriched assessment. *Research & Practice in Assessment, 9*. 5–16. Retrieved from http://www.rpajournal.com/the-future-of-data-enriched-assessment/

---

[13] https://www.csc2.ncsu.edu/faculty/healey/tweet_viz/tweet_app/

8. Wellings, J., & Levine, M. H. (2009). *The digital promise: transforming learning with innovative uses of technology*. Sesame Workshop.
9. Romero, C., & Ventura, S. (2013). Data mining in education. *Wiley Interdisciplinary Reviews: Data Mining and Knowledge Discovery, 2013*(3), 12–27. https://doi.org/10.1002/widm.1075.
10. Romero, C., & Ventura, S. (2010). Educational data mining: A review of the state of the art. *IEEE Transactions on Systems, Man, and Cybernetics, Part C: Applications and Reviews, 40*(6), 601–618.
11. Huang, S., & Fang, N. (2013). Predicting student academic performance in an engineering dynamics course: A comparison of four types of predictive mathematical models. *Computers & Education, 61*, 133–145.
12. Romero-Zaldivar, V.-A., Pardo, A., Burgos, D., & Kloos, C. D. (2012). Monitoring student progress using virtual appliances: a case study. *Computers & Education, 58*(4), 1058–1067.
13. Parry, M. (2010). Like Netflix, New College Software Seeks to Personalize Recommendations, The chronicle of Higher Education. Available from http://chronicle.com/blogs/wiredcampus/likenetflix-new-college-software-aims-to-personalize-recommendations/27642
14. Kop, R. (2010, June). The design and development of a personal learning environment: Researching the learning experience, European Distance and E-learning Network annual Conference 2010.
15. Valencia, Spain, Paper H4 32. (2010). Laat, de, M.: Networked learning, PhD thesis, Instructional Science, Utrecht Universiteit, The Netherlands (2006).
16. Siemens, G., & Baker, R. S. J. D. (2012). Learning analytics and educational data mining: towards communication and collaboration. In *Proceedings of the 2nd International Conference on Learning Analytics and Knowledge* (pp. 1–3). British Columbia, Canada: Vancouver.
17. Baker, R. S. J. D., Gowda, S. M., & Corbett, A. T. (2011). Automatically detecting a student's preparation for future learning: help use is key. In *Fourth International conference on educational data mining* (pp. 179–188). The Netherlands: Eindhoven.
18. Romero, C., Espejo, P., Zafra, A., Romero, J., & Ventura, S. (2013). Web usage mining for predicting marks of students that use Moodle courses. *Computer Applications in Engineering Education, 21*(1), 135–146.
19. Vellido, A., Castro, F., & Nebot, A. (2011). *Clustering educational data. Handbook of educational data mining* (pp. 75–92). Boca Raton: Chapman and Hall/CRC Press.
20. Amershi, S., & Conati, C. (2009). Combining unsupervised and supervised classification to build user models for exploratory learning environments. *Journal of Educational Data Mining, 1*(1), 18–71.
21. Anaya, A. R., & Boticario, J. G. (2009). A data mining approach to reveal representative collaboration indicators in open collaboration frameworks. In T. Barnes, M. Desmarais, C. Romero, & S. Ventura (Eds.), *Educational data mining 2009: Proceedings of the 2nd International conference on educational data mining* (pp. 210–219).
22. Ueno, M. (2004). Online outlier detection system for learning time data in e-learning and its evaluation. In *International Conference on Computers and Advanced Technology in Education. Beijiing, China* (pp. 248–253).
23. Merceron, A., & Yacef, K. (2010). Measuring correlation of strong symmetric association rules in educational data. In C. Romero, S. Ventura, M. Pechenizkiy, & R. S. J. D. Baker (Eds.), *Handbook of educational data mining* (pp. 245–256). Boca Raton, CRC Press.
24. Rabbany, R., & Takaffoli, M. (2011). Zäiane O. Analyzing participation of students in online courses using social network analysis techniques. In *International conference on educational data mining* (pp. 21–30). The Netherlands: Eindhoven.
25. Trcka, N., Pechenizkiy, M., & van der Aalst, W. (2011). Process mining from educational data. In *Handbook of educational data mining* (pp. 123–142). Boca Raton: CRC Press.
26. Tane, J., Schmitz, C., & Stumme, G. (2004). Semantic resource management for the web: An e-learning application. In: *International Conference of the WWW*. New York, pp. 1–10.
27. Mazza, R., & Milani, C. (2004). GISMO: A graphical interactive student monitoring tool for course management systems. In: *International conference on technology enhanced learning*. Milan, Italy, pp. 1–8.

28. Baker, R. S. J. D., & Yacef, K. The state of educational data mining in 2009: A review and future visions. *J Edu Data Min, 2009*, 3–17.
29. Bienkowski, M., Feng, M., & Means, B. (2012). *Enhancing teaching and learning through educational data mining and learning analytics: an issue brief* (pp. 1–57). Washington, DC: Office of Educational Technology, U.S. Department of Education.
30. Johnson, L., et al. (2016). *NMC horizon report: 2016 higher education edition*. Austin: The New Media Consortium.
31. Shacklock, X. (2016). *From bricks to clicks: The potential of data and analytics in higher education*. Higher Education, Committee.
32. Papamitsiou, Z., & Economides, A. (2014). Learning analytics and educational data mining in practice: A systematic literature review of empirical evidence. *Educational Technology & Society, 17*(4), 49–64.
33. Abdous, M., He, W., & Yen, C.-J. (2012). Using data mining for predicting relationships between online question theme and final grade. *Educational Technology & Society, 15*(3), 77–88.
34. Lykourentzou, I., Giannoukos, I., Nikolopoulos, V., Mpardis, G., & Loumos, V. Dropout prediction in e-learning courses through the combination of machine learning techniques. *Computer & Education, 53*, 950–965.
35. Dekker, G. W., Pechenizkiy, M., & Vleeshouwers, J. M. (2009). Predicting students drop out: A case study. In T. Barnes, M. Desmarais, C. Romero, & S. Ventura (Eds.), *Proceedings of the 2nd International conference on educational data mining* (pp. 41–50). Retrieved from http:// www.educationaldatamining.org/EDM2009/uploads/proceedings/dekker.pdf
36. Kizilcec, R. F., Piech, C., & Schneider, E. (2013). Deconstructing disengagement: Analyzing learner subpopulations in massive open online courses. In D. Suthers, K. Verbert, E. Duval, & X. Ochoa (Eds.), *Proceedings of the 3rd International conference on learning analytics and knowledge* (pp. 170–179). New York, NY: ACM.
37. Giesbers, B., Rienties, B., Tempelaar, D., & Gijselaers, W. (2013). Investigating the relations between motivation, tool use, participation, and performance in an e-learning course using web-videoconferencing. *Computers in Human Behavior, 29*(1), 285–292.
38. He, W. (2013). Examining students' online interaction in a live video streaming environment using data mining and text mining. *Computers in Human Behavior, 29*(1), 90–102.
39. Dejaeger, K., Goethals, F., Giangreco, A., Mola, L., & Baesens, B. (2012). Gaining insight into student satisfaction using comprehensible data mining techniques. *European Journal of Operational Research, 218*(2), 548–562.
40. Xing, W., Guo, R., Petakovic, E., & Goggins, S. (2015). Participation-based student final performance prediction model through interpretable Genetic Programming: Integrating learning analytics, educational data mining and theory. *Computers in Human Behavior, 47*, 168–181. https://doi.org/10.1016/j.chb.2014.09.034.
41. Zacharis, N. Z. (2015). A multivariate approach to predicting student outcomes in web-enabled blended learning courses. *Internet and Higher Education, 27*, 44–53. https://doi.org/10.1016/j.iheduc.2015.05.002.
42. Niemi, D., & Gitin, E. (2012). *Using big data to predict student dropouts: Technology affordances for research*. International Association for Development of the Information Society, Paper presented at the International Association for Development of the Information Society (IADIS) International Conference on Cognition and Exploratory Learning in Digital Age (CELDA) (Madrid, Spain, Oct 19–21, 2012), 4 pp.
43. Junco, R., & Clem, C. (2015). Predicting course outcomes with digital textbook usage data. *Internet & Higher Education, 27*, 54–63. https://doi.org/10.1016/j.iheduc.2015.06.001.
44. Mouri, K., Okubo, F., Shimada, A., & Ogata, H. (2016). *Bayesian Network for predicting students' final grade using e-book Logs in University Education*. In Proceedings of 16th international conference on Advanced learning technologies, July 2016, https://doi.org/10.1109/ICALT.2016.27

45. Blikstein, P. (2011). Using learning analytics to assess student' behavior in open-ended programming tasks. In P. Long, G. Siemens, G. Conole, & D. Gasevic (Eds.), *Proceedings of the 1st International conference on learning analytics and knowledge* (pp. 110–116). New York, NY: ACM.
46. Moridis, C. N., & Economides, A. A. (2009). Prediction of student's mood during an online test using formula-based and neural network-based method. *Computers & Education, 53*(3), 644–652.
47. Jeong, H., Choi, C., & Song, Y. (2012). Personalized learning course planner with e-learning DSS using user profile. *Expert Systems with Applications, 39*(3), 2567–2577.
48. Macro, A., Agnes, K. H., Inmaculada, A. S., & Gábor, K. (2012). Meta-analyses from a collaborative project in mobile lifelong learning. *British Educational Research Journal, 20,* 1), 1–1),26.
49. Lin, C. F., Yeh, Y.-c., Hsin Hung, Y., & Chang, R. I. (2013). Data mining for providing a personalized learning path in creativity: An application of decision trees. *Computers & Education, 68*(2013), 199–210.
50. Hsu, M. (2008). A personalized English learning recommender system for ESL students. *Expert Systems with Applications, 34*(1), 683–688.
51. Méndez, J. R., Fdez-Riverola, F., Iglesias, E. L., Díaz, F., & Corchado, J. M. (2006). Tracking concept drift at feature selection stage in Spam Hunting: An anti-spam instance-based reasoning system. *Lecture Notes in Computer Science, 4106*, 504–518.
52. Khaing, K. T. (2010). Enhanced features ranking and selection using recursive feature elimination (RFE) and K-Nearest Neighbor algorithms. *International Journal of Network and Mobile Technologies, 1*(1), 1–12.
53. Chien, C., Wang, W., & Cheng, J. (2007). Data mining for yield enhancement in semiconductor manufacturing and an empirical study. *Expert Systems with Applications, 33*(1), 192–198.
54. Lee, Y.-J. (2012). Developing an efficient computational method that estimates the ability of students in a Web-based learning environment. *Computers & Education, 58*(2012), 579–589.
55. Shute, V. J. (2008). Focus on formative feedback. *Review of Educational Research, 78*(1), 153–189. https://doi.org/10.3102/0034654307313795.
56. Romero, C., Zafra, A., Luna, J. M., & Ventura, S. (2013). Association rule mining using genetic programming to provide feedback to instructors from multiple-choice quiz data. *Expert Systems, 30*(2), 162–173.
57. Pechenizkiy, M., Calders, T., Vasilyeva, E., & De Bra, P. (2008). Mining the student assessment data: Lessons drawn from a small scale case study, *International Conference on Educational Data Mining*, Cordoba, Spain, pp. 187–191
58. Kizilcec, R. F., Bailenson, J. N., & Gomez, C. J. (2015). The instructor's face in video instruction: Evidence from two large-scale field studies. *Journal of Educational Psychology, 107*(3), 724–739.
59. Kizilcec, R. F., & Brooks, C. (2017). Diverse big data and randomized field experiments in massive open online courses: Opportunities for advancing learning research. In G. Siemens & C. Lang (Eds.), *Handbook on learning analytics & educational data mining.* New York: Springer.
60. Baker, R., Dee, T., Evans, B., & John, J. (2015). *Bias in online classes: Evidence from a field experiment.* Paper presented at the SREE Spring 2015 Conference, Learning Curves: Creating and Sustaining Gains from Early Childhood through Adulthood, 5–7 March 2015, Washington, DC, USA.
61. Kizilcec, R. F., Pérez-Sanagustín, M., & Maldonado, J. J. (2016). *Recommending self-regulated learning strategies does not improve performance in a MOOC.* Proceedings of the 3rd ACM Conference on Learning @ Scale (L@S 2016), 25–28 April 2016, Edinburgh, Scotland (pp. 101–104). New York: ACM.
62. Rogers, T., & Feller, A. (2016). Discouraged by peer excellence: Exposure to exemplary peer performance causes quitting. *Psychological Science, 27*(3), 365–374.

63. Romero, C. Ventura, S. De Bra, P., & De Castro, C. (2002). *Discovering prediction rules in AHA! courses.* In: 9th International Conference on User Modeling, Johnstown, PA, USA, pp. 25–34.

64. Romero, C., Ventura, S., & De Bra, P. (2004). Knowledge discovery with genetic programming for providing feedback to courseware author. *User Model User-Adapted Interact, 14*, 425–464.

65. Benchaffai, M., Debord, G., Merceron, A., & Yacef, K.. (2004). *TADA-Ed, a tool to visualize and mine students' online work.* In B. Collis (Eds.), Proceedings of International conference on computers in education, (ICCE04) (pp 1891–1897). Melbourne, Australia: RMIT.

66. Merceron, A., & Yacef, K. (2005). TADA-Ed for educational data mining, interactive multimedia electronic. *Journal of Computer-Enhanced Learning, 7*(1), http://imej.wfu.edu/articles/2005/1/03/index.asp

67. Avouris, N., Komis, V., Fiotakis, G., Margaritis, M., & Voyiatzaki, E.. (2005). Why logging of fingertip actions is not enough for analysis of learning activities. In: *Workshop on usage analysis in learning systems.* AIED Conference: Amsterdam, pp. 1–8.

68. Avouris, N., Fiotakis, G., Kahrimanis, G., Margaritis, M. & Komis, V. (2007). Beyond logging of fingertip actions: analysis of collaborative learning using multiple sources of data. *Journal of Interactive Learning Research, Association for the Advancement of Computing in Education, 18*(2) Special Issue: Usage Analysis in Learning Systems : Existing Approaches and Scientific Issues, pp. 231–250.

69. Jovanović, J., Gašević, D., Brooks, C. A., Eap, T., Devedžić, V., Hatala, M., & Richards, G. (2008). LOCO-analyst: Semantic web technologies to analyze the use of learning content. *International Journal of Continuing Engineering Education and Life-Long Learning, 18*(1), 54–76.

70. Ali, L., Hatala, M., Gasevic, D., & Jovanovic, J. (2012). A qualitative evaluation of evolution of a learning analytics tool. *Computers & Education, 58*, 470–489.

71. Koedinger, K., Cunningham, K., Skogsholm, A., (2008). Leber, B. An open repository and analysis tools for fine-grained, longitudinal learner data. In: *First International conference on educational data mining.* Montreal, Canada, pp. 157–166.

72. Koedinger, K. R., Baker, R. S. J. D., Cunningham, K., Skogsholm, A., Leber, B., & Stamper, J. (2010). A data repository for the EDM community: The PSLC DataShop. In C. Romero, S. Ventura, M. Pechenizkiy, & R. S. J. D. Baker (Eds.), *Handbook of educational data mining.* Boca Raton: CRC Press.

73. Rummel, N., Spada, H., & Diziol, D. (2007). *Evaluating collaborative extensions to the Cognitive Tutor Algebra in an in vivo experiment.* Lessons learned. Paper presented at the 12th European Conference for Research on Learning and Instruction (EARLI). Budapest, Hungary.

74. Hausmann, R., & VanLehn, K. (2007). Self-explaining in the classroom: Learning curve evidence. In McNamara & Trafton (Eds.), *Proceedings of the 29th Annual Cognitive Science Society* (pp. 1067–1072). Austin, TX: Cognitive Science Society.

75. McLaren, B. M., Lim, S., Yaron, D., & Koedinger, K. R. (2007). Can a polite intelligent tutoring system lead to improved learning outside of the lab? In Luckin & Koedinger (Eds.), *Proceedings of the 13th International Conference on Artificial Intelligence in Education* (pp. 433–440). Los Angeles: IOS Press.

76. Pavlik Jr., P. I., Presson, N., & Koedinger, K. R. (2007). Optimizing knowledge component learning using a dynamic structural model of practice. In R. Lewis, & T. Polk (Eds.), Proceedings of the Eighth International Conference of Cognitive Modeling.

77. Garcia, E., Romero, C., Ventura, S., & Castro, C. (2009). Collaborative data mining tool for education. In *International Conference on Educational Data Mining.* Cordoba, Spain, pp. 299–306.

78. Juan, A., Daradoumis, T., Faulin, J., & Xhafa, F. (2009). SAMOS: a model for monitoring students' and groups' activities in collaborative e-learning. *International Journal of Learning Technology, 4*, 53–72.

79. Gaudioso, E., Montero, M., Talavera, L., & Hernandez-del-Olmo, F. (2009). Supporting teachers in collaborative student modeling: A framework and an implementation. *Expert Systems with Applications, 36*, 2260–2265.

80. Rabbany, R., El Atia, S, Takaffoli, M., & Zaiane, O. R. (2013). Collaborative learning of students in online discussion forums: A social network analysis perspective. In the Springer edited book Educational Data Mining: Applications and Trends, Springer Series: Studies in Computational Intelligence.
81. Matsuzaw, Y., Oshima, J., Oshima, R., Niihara, Y., & Sakai, S. (2011). KBDeX: A platform for exploring discourse in collaborative learning. *Procedia-Social and Behavioral Sciences, 26*, 198–207.
82. Matsuzawa, Y., Oshima, J., Oshima, R., & Sakai, S. (2012). Learners' use of SNA-based discourse analysis as a self-assessment tool for collaboration. *International Journal of Organisational Design and Engineering, 2*(4), 362–379.
83. Luna, J. M., Castro, C., & Romero, C. (2017). MDM tool: A data mining framework integrated into Moodle. *Computer Applications in Engineering Education, 25*(1), 90–102.
84. Graf, S., Ives, C., Rahman, N., & Ferri, A. (2011). AAT-A tool for accessing and analysing student's behaviour data in learning systems. In: *Proceedings of the 1st International Conference on Learning Analytics and Knowledge*, Lak, Banff, AB, Canada, pp. 174–179.
85. Dragulescu, B., Bucos, M., & Vasiu, R. (2015). *CVLA: Integrating multiple analytics techniques in a custom Moodle report.* International Conference ICIST, Druskininkai, Lithuania, pp. 115–126.
86. Garcıa-Saiz, D., & Zorrilla, M. E. (2013). A service oriented architecture to provide data mining services for non-expert data miners. *Decision Support Systems, 55*, 399–411.
87. Mazza, R., Bettoni, M., Fare, M., & Mazzola, L. (2012). *MOCLog—Monitoring online courses with log data.* 1st Moodle Research Conference Proceedings, Heraklion, Greece, pp. 132–139.
88. Bakharia, A., & Dawson, S. (2011). SNAPP: A bird's-eye view of temporal participant interaction. International conference on learning analytics and knowledge. ACM, New York, NY, USA, pp. 168–173.
89. Jakub Kuzilek et al. (2015, March). 'OU Analyse: Analysing at-risk students at The Open University. *Learning analytics community exchange learning analytics review.*
90. Sundorph, E., & Mosseri-Marlio, W. (2016, September). *Smart campuses: How big data will transform higher education.* Retrieved from www.reform.uk/wp-content/uploads/2016/09/Smart-campuses-WEB.pdf
91. CASE STUDY I: Predictive analytics at Nottingham Trent University. (2016). Retrieved from https://analytics.jiscinvolve.org/wp/files/2016/04/CASE-STUDY-I-Nottingham-Trent-University.pdf
92. Bhandari, G., & Gowing, M. (2016). *A framework for open assurance of learning.* Proceedings of the 12th International Symposium on Open Collaboration, August 17–19, 2016.
93. Bail'on, M., Carballo, M., Cobo, C., Magnone, S., Marconi, C., Mateu, M., & Susunday, H. (2015). How can plan Ceibal land into the age of big data? *In 4th International Conference on Data Analytics*, pp. 126–129.
94. Aguerrebere, C., Cobo, C., Gomez, M., Mateu, M. (2017). Strategies for data and learning analytics informed national education policies: The Case of Uruguay. In *Proceedings of the seventh international learning analytics & knowledge conference* (pp. 449–453). Vancouver, British Columbia, Canada, March 13–17, 2017.
95. Kim, D. R., Hue, J.-P., & Shin, S.-S. (2016, December). Application of learning analytics in University Mathematics Education. *Indian Journal of Science and Technology, 9*(46). https://doi.org/10.17485/ijst/2016/v9i46/107193.
96. Ray, S. K., & Saeed, M. Mobile learning using social media platforms: An empirical analysis of users' behaviors. *International Journal of Mobile Learning and Organization, 6*(3), 258–270. Inderscience publications, 2015.
97. OECD. (2013). *OECD report: The State of Higher Education 2013.* http://www.oecd.org/edu/imhe/thestateofhighereducation2013.htm
98. Dede, C. (2016). Next steps for "Big Data" in education: Utilizing data-intensive research. *Educational Technology, LVI*(2), 37–42.

Chapter 8
Handling Pregel's Limits in Big Graph Processing in the Presence of High-Degree Vertices

Mohamad Al Hajj Hassan and Mostafa Bamha

8.1 Introduction

Graphs are widely used for data representation in many domains such as road networks, social networks, bioinformatics, and computer networks. Due to this wide adoption of graphs, graph mining has received large interest from researchers. Graph mining allows users to analyze and have deeper understanding of the graph data in order to extract meaningful information and nontrivial patterns [4]. Connected components, graph clustering, page rank, and node proximities are among the graph mining algorithms proposed in the literature. These algorithms are used in many applications such as community detection, link prediction, and influence propagation in social media [19] and anomaly detection [2]. These algorithms assume that the graph fits in the memory of the processing machine [12, 18]. However today with the exponential increase in the volume of data collected by organizations, this assumption is not valid in many applications. To this end, distributed processing of these mining algorithms, known as big graph mining, is increasingly used to handle this problem [4, 5]. However, most of these graph analysis algorithms are iterative. For this reason, general-purpose distributed data

This work was started when the author was working at Lebanese International University, Lebanon.

M. Al Hajj Hassan (✉)
Huawei, German Research Center, Munich, Germany
e-mail: mohamad.alhajjhassan@huawei.com

M. Bamha
INSA Centre Val de Loire, Université Orléans, Orléans, France
e-mail: Mostafa.Bamha@univ-orleans.fr

© Springer International Publishing AG, part of Springer Nature 2018

M. M. Alani et al. (eds.), *Applications of Big Data Analytics*,
https://doi.org/10.1007/978-3-319-76472-6_8

processing frameworks such as Hadoop [3] are not efficient for processing such algorithms. This is mainly because, for each iteration, input graph data should be read from the Distributed File System (DFS) and the partial output should be written back to the DFS which may induce high disk I/O in big graph processing in large-scale systems.

Several specialized graph processing frameworks such as Google's Pregel [15], Apache Giraph [8], GraphLab [14], and PowerGraph [9] are proposed to speed up the execution of iterative graph algorithms. Most of these frameworks follow *"think like a vertex"*: a vertex-centric programming model. In frameworks such as Google's Pregel and Giraph, based on bulk synchronous parallel (BSP) model [20], each vertex receives messages from its incoming neighbors, updates its state, and then sends messages to outgoing neighbors in each iteration. On the other side, PowerGraph (an improved version of GraphLab) is based on GAS (Gather, Apply, Scatter) model and a shared memory abstraction. In this model, each active node can directly access and collect data from its neighbors, in the Gather phase, without the need for messages. Each active vertex then accumulates the collected values to update its state in the Apply phase then updates and activates its adjacent vertices in the scatter phase.

In all these frameworks, graph vertices are partitioned into subgraphs which are distributed over a set of computing nodes; however graph partitioning is considered as NP-complete problem [7]. In order to benefit from the processing capacity of parallel and distributed machines, the partitions should be assigned to computing nodes in a way that balances their workload and reduces communication costs. Due to power-law degree distribution in real graphs where a few number of vertices are connected to a large fraction of the graph [1, 13] such as celebrities on Facebook and Twitter, the performance of Pregel-like systems degrades in the presence of high-degree vertices in large-scale systems [21]. So partitioning of these high-degree vertices must be performed (using techniques like vertex mirroring/splitting) to achieve good performance. To this end, we propose a MapReduce-based graph partitioning algorithm that allows us to evenly assign the load to all computing nodes. The algorithm proceeds in two MapReduce jobs. In the first one, the graph is read from HDFS[1] to identify high-degree vertices and to define the appropriate number of slave vertices to balance processing time among workers. In the second job, we create for each high-degree vertex a master vertex in addition to incoming and outgoing slave vertices. Slave vertices are evenly assigned to distinct workers in order to balance their loads. Graph analysis algorithms can be applied on the partitioned graph using any framework that supports reading data from HDFS. We tested the performance of our approach by executing *Single Source Shortest Paths* (SSSP) algorithm on partitioned and unpartitioned graphs with highly skewed nodes

---

[1] HDFS: Hadoop Distributed File System.

under Hadoop and Giraph frameworks. The test results proved the efficiency of using our graph partitioning approach as a preprocessing step to make Pregel-like systems scalable.

The remaining of the article is organized as follows. In Sect. 8.2, we review MapReduce and Pregel programming models. Our big graph partitioning and processing approach is presented in Sect. 8.3 with its complexity analysis. Experiment results presented in Sect. 8.4 confirm the efficiency of our approach. Related works are reviewed in Sect. 8.5; we then conclude in Sect. 8.6.

8.2   MapReduce vs Pregel Programming Model

Google's MapReduce programming model presented in [6] is based on two functions: *map* and *reduce*. Dean and Ghemawat stated that they have inspired their MapReduce model from Lisp and other functional languages [6]. The programmer is only required to implement two functions **map** and **reduce** having the following signatures:

map: $(k_1, v_1) \longrightarrow list(k_2, v_2),$
reduce: $(k_2, list(v_2)) \longrightarrow list(v_3).$

The user must write the **map** function that has two input parameters, a key k_1 and an associated value v_1. Its output is a list of intermediate key/value pairs (k_2, v_2). This list is partitioned by the MapReduce framework depending on the values of k_2, where all pairs having the same value of k_2 belong to the same group and all key-value pairs corresponding to the same group are sent to the same reducer.

The **reduce** function that must also be written by the user has two parameters as input: an intermediate key k_2 and a list of intermediate values $list(v_2)$ associated with k_2. It applies the user-defined merge logic on $list(v_2)$ and outputs a list of values $list(v_3)$.

MapReduce is a simple yet powerful framework for implementing distributed applications without having extensive prior knowledge of issues related to data redistribution or task allocation and fault tolerance in large-scale distributed systems. Most MapReduce frameworks include *Distributed File Systems* (DFS) designed to store very large files with streaming data access patterns and data replication for fault tolerance while guaranteeing high disk I/O throughput.

To cover a large set of application needs in term of computation and data redistribution, in most MapReduce frameworks, the user can optionally implement two additional functions : init() and close() called before and after each map or reduce task. The user can also specify a **partition** function to send each key-value pair (k_2, v_2), generated in map phase, to a specific reducer. Destination reducer may be computed using only a part of the input key k_2 (Hadoop's default "partition function" is generally based on "hashing" the whole input key k_2) [10, 11]. The signature of the partition function is :

partition: $(Key \quad k_2) \longrightarrow Integer.$ `/* Integer is between 0 and the`
`                                  number of reducers #numReduceTasks */`

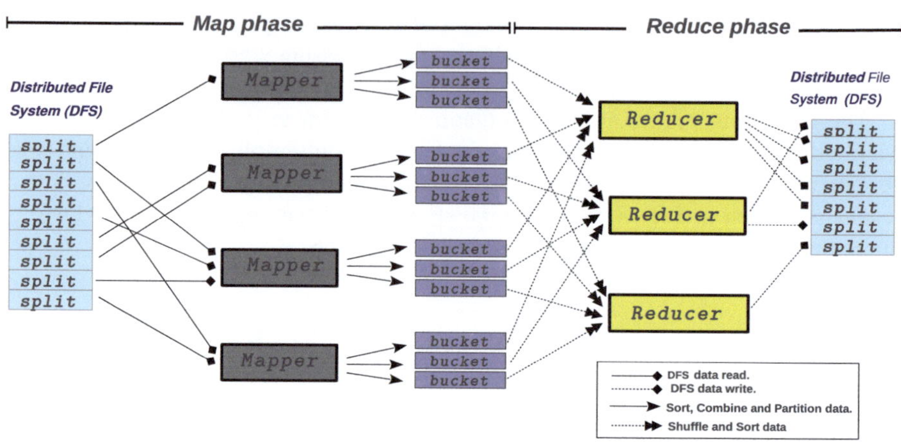

Fig. 8.1 MapReduce framework

For efficiency reasons, in Hadoop MapReduce framework, the user may also specify a **combine** function to reduce the amount of data transmitted from Mappers to Reducers during *shuffle* phase (see Fig. 8.1). The "combine function" is like a local reduce applied (at map worker) before storing or sending intermediate results to the reducers. The signature of **combine** function is:

combine: $(k_2, list(v_2)) \longrightarrow (k_2, list(v_3))$.

MapReduce excels in the treatment of data parallel applications where computation can be decomposed into many independent tasks involving large input data. However MapReduce performance degrades in the case of dependent tasks or iterative data processing such as graph computations due to the fact that, for each computation step, input data must be read from DFS for each map phase and intermediate output data must be written back to DFS at the end of reduce phase for each iteration : this may induce high communication and disk I/O costs. To this end, *Pregel* model (a variant of MapReduce) [15] was introduced for large-scale graph processing. Pregel is based on *bulk synchronous parallel* (BSP) programming model [20] where each parallel program is executed as a sequence of parallel *supersteps*. Each *superstep* is divided into (at most) three successive and logically disjoint phases. In the first phase, each processing node uses its local data (only) to perform sequential computations and to request data transfers to/from other nodes. In the second phase, the network delivers the requested data transfers, and in the third phase, a global synchronization barrier occurs making the transferred data available for the next superstep.

To minimize execution time of a BSP program, its design must jointly minimize the number of supersteps and the total volume of communication while avoiding load imbalance among processing nodes during all the stages of computation.

Similarly to BSP, Pregel programs proceed into three logical phases:

1. A setup phase orchestrated by a master node where *workers* read input graph data from DFS. The graph vertices are partitioned among these *workers* using a partitioning function applied on vertices identifiers. The partitioning is based on hashing functions, and each partition is assigned to a single worker,
2. A processing phase where each worker performs a sequence of supersteps. Each superstep consists of a *local computation* (defined in compute() function) on worker's data: active vertices can modify their state or that of its outgoing edges and handle received messages sent to them in the previous superstep. This step is followed by a *communication* step where each worker can send asynchronously messages or vertices/edges modification requests to other processing nodes (these messages are received in the following superstep to perform vertex data/state modification, Vertex/Edges modifications, ...), and then a *global synchronization barrier* is performed to make transferred data and graph topology modifications available for the next superstep,
3. A close phase where workers store output result to DFS[2] and stop graph processing.

Figure 8.2 shows an example of a *superstep* computation in Pregel-like systems where only a set of "active" vertices performs a local computation. Each idle vertex is activated whenever it receives one or more messages from other vertices. The program terminates (halts) when all vertices become inactive and there is no pending message in the following superstep.

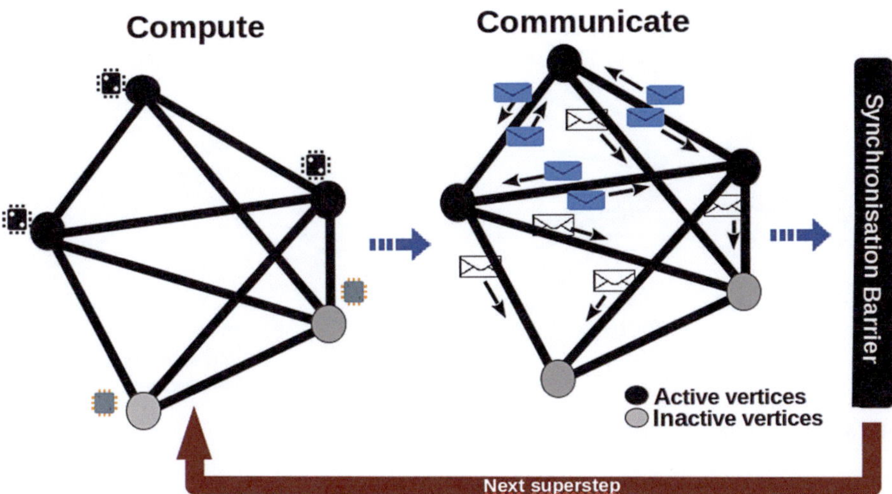

Fig. 8.2 Supersteps in Pregel computation model for a graph processing

---

[2]DFS: Distributed File System.

Even if Pregel scales better than MapReduce for graph processing, it still remains inefficient in the presence of high-degree vertices since for each processing iteration, high-degree vertices may communicate with all their neighbors which can induce a load imbalance among processing nodes. This can also lead to a memory lack whenever these messages or the list of neighbors cannot fit in processing node's memory which limits the scalability of the model. To this end, we introduce, in this article, a partitioning approach for high-degree vertices based on a master/slave repartition allowing to avoid the load imbalance among processing nodes while guaranteeing that the amount of data communicated at each computation step never exceeds a user-defined value. This partitioning can be seen as a graph preprocessing step before using Pregel-like systems to make most of existing graph algorithms scalable by applying some minor changes to original graph algorithms.

In this partitioning, a high-degree vertex **H** is transformed into a master vertex called **H-0** connected to set of "left" and "right" slaves (called $H - L_{i=1..m}$ and $H - R_{j=1..n}$, respectively) depending on the number of the incoming and outgoing edges to vertex **H**. Figure 8.3 shows an example where a high-degree vertex, **H**, is partitioned into m "left" slaves and n "right" slaves. *Left* and *right* slaves are then affected to distinct workers, in a round-robin manner, to balance computation and communication in each graph processing step in Pregel-like systems.

In this article, we used an open-source version of MapReduce called Hadoop developed by "The Apache Software Foundation". Hadoop framework includes a Distributed File System called HDFS designed to store very large files with streaming data access patterns.

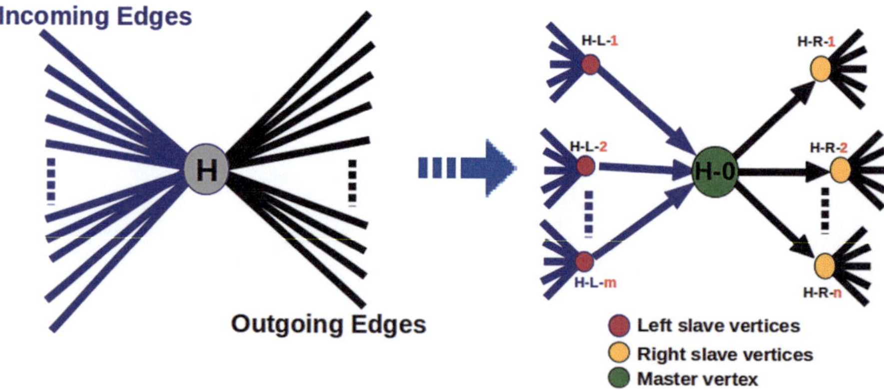

Fig. 8.3 High degree vertex partitioning approach

8.3 GPHDV: A Solution for Big Graph Partitioning/Processing in the Presence of High-Degree Vertices

We introduce, in this section, a new approach called *GPHDV* (graph partitioning for high-degree vertices) used to avoid the effect of high-degree vertices in graph processing. Our approach evenly partitions each high-degree vertex into a master and a set of left and right slave vertices depending on the number of incoming and outgoing edges to each high-degree vertex. This partitioning is used as a preprocessing phase to make Pregel-like systems scalable while guaranteeing perfect balancing properties of communication and computation during all stages of big graph processing. Vertex partitioning, in this approach, can be generalized to many graph processing problems ranging from *SSSP*, *PageRank* to *Connected Components* problems in all Pregel-like systems.

We will describe, in details, the computation steps of *GPHDV* while giving an upper bound of execution cost for each step. The O(...) notation only hides small constant factors: they only depend on program's implementation but neither on data nor on machine parameters. GPHDV proceeds in two MapReduce jobs:

a. the first job is used to identify high degree vertices and to generate high degree vertices partitioning templates,
b. the second job is used to partition *input graph* data using generated partitioning templates. In this partitioning, only high degree vertices are partitioned into masters and slave vertices. These slave vertices are affected to different workers in a *round-robin* manner to balance load among processing nodes.

We consider a weighted n-vertex, m-edge graph G(V, E) where V is the set of vertices of G and E the set of edges in G. For scalability, we assume that graph G(V, E) is stored as a set of edges E divided into blocks (splits) of data. These splits are stored in Hadoop Distributed File System (HDFS). They are also replicated on several nodes for reliability issues. Throughout this article, we use the following notations:

- $|E|$: the number of pages (or blocks of data) forming the set of edges E,
- $\|E\|$: the number m of edges in E,
- E_i^{map}: the split(s) of set E affected to mapper (Worker) i,
- E_i^{red}: the split(s) of set E affected to reducer (Worker) i,
- $|V|$: number of pages (or blocks of data) forming the set of vertices V
- $\|V\|$: number n of vertices in V
- $\overline{V}$: the restriction (a fragment) of set V which contains only high-degree vertices. $\|\overline{V}\|$ is, in general, very small compared to the number of vertices $\|V\|$,
- V_i^{map}: the split(s) of set V affected to mapper (Worker) i, V_i^{map} contains all (source and destination) vertices of E_i^{map},
- V_i^{red}: the split(s) of set V affected to reducer (Worker) i. V_i^{red} is the subset of V obtained by a simple hashing of vertices in V,

- $\overline{V_i}$: the split(s) of set $\overline{V}$ affected to mapper i holding only high-degree vertices from V,
- $c_{r/w}$: read/write cost of a page of data from/to Distributed File System (DFS)
- c_{comm}: communication cost per page of data,
- t_s^i: time to perform a simple search in a Hashtable on node i,
- t_h^i: time to add an entry to a Hashtable on node i,
- NB_mappers: number of job mapper nodes,
- NB_reducers: number of job reducer nodes.

GPHDV proceeds into three phases (two phases are carried out in the first MapReduce Job and a single map phase in the second Job) :

Algorithm 1 GPHDV algorithm workflow

---

a.1 ▶ Map phase of the 1st MapReduce job:
　　/* To generate "local" incoming and outgoing degrees for each vertex in the input graph */
▷ Each mapper i reads its assigned data splits (blocks) of subset E_i^{map} from the DFS
　　▷ Get `source vertex` "s" and `destination vertex` "d" from each input edge
　　　e(source vertex : s, destination vertex : d, value : a) from E_i^{map}.
　　▷ Emit a couple ($<$A,s$>$,1) /* Tag "A" to identify an outgoing edge from source
　　　vertex "s" */
　　　▷ Emit a couple ($<$B,d$>$,1) /* Tag "B" to identify an incoming edge to vertex "d" */
▶ Combine phase: /* To compute local frequencies (incoming and outgoing degrees) for each
source and destination vertex in set E_i^{map} */
▷ Each combiner, for each source vertex "s" (resp. destination vertex "d") computes local
outgoing (resp. local incoming) degree : the sum of generated local frequencies associated to
a source (resp. destination) vertex generated in Map phase.
▶ Partition phase:
▷ For each emitted couple (key, value)=($<$Tag,"v"$>$,frequency) where Tag is "A" for (resp. "B")
source (resp. destination) vertex "v", compute reducer destination according to only vertex ID "v".
a.2 ▶ Reduce phase: /* To combine Shuffle's records and to create Global histogram for
　　　　　　　　　　　　　　　　"high-degree vertices partitioning templates" */
▷ Compute the global frequencies (incoming and outgoing degrees) for each vertex present in
　set V
▷ Emit, for each high-degree vertex "v", a couple ("v",NbLeftSlaveVertices,
　NbRightSlaveVertices):
`NbLeftSlaveVertices` and `NbRightSlaveVertices` are fixed depending on the values
of the incoming and outgoing degrees of vertex "v".
b.1 ▶ Map phase of the 2nd MapReduce job:
▷ Each mapper reads Global histogram of "high-degree vertices partitioning templates" from
　DFS
and creates a local Hashtable.
▷ Each mapper, i, reads its assigned edge splits of input graph from DFS and generates a set
　of left and right slave vertices for each edge depending on the degrees of source and
destination vertices: only edges associated to a high degree for a source or a destination vertex are
transformed whereas those associated to low degree for both source and destination vertices are
emitted as they are without any graph transformation. Note that, left and right slave vertices
associated to a high degree vertex are created only once by a designated worker and new edges
are created from each master vertex to its left and right slave vertices.
▷ Emit all edges resulting from this graph transformation and mark high-degree vertices for a
later removal.

---

a.1: Map phase to generate mapper's "local" incoming and outgoing degrees for each vertex in mapper's input graph

Each mapper i reads its assigned data splits (blocks) of subset E_i^{map} from the DFS. It extracts the `source vertex` "s" and the `destination vertex` "d" from each input edge e(source vertex : s, destination vertex : d, value : a) and emits two tagged records for "s" and "d" vertices with frequency "One" as follows:

- Emit a couple ($<$A,s$>$,1): This means that there exists "One" outgoing edge starting from `source vertex` "s."
- Emit a couple ($<$B,d$>$,1): This means that there exists "One" incoming edge arriving to `destination vertex` "d."

Tag "A" is used to identify outgoing edges from a source vertex, whereas tag "B" is used to identify incoming edges to a destination vertex. The cost of this step is at most:

$$Time(a.1.1) = O\left(\max_{i=1}^{NB\_mappers} c_{r/w} * |E_i^{map}| + 2 * \max_{i=1}^{NB\_mappers} \|E_i^{map}\| \right).$$

The term $\max_{i=1}^{NB\_mappers} c_{r/w} * |E_i^{map}|$ is time to read input graph data from HDFS by all the mappers, whereas the term $2*\max_{i=1}^{NB\_mappers} \|E_i^{map}\|$ is time to scan mapper's edges and to emit two tagged records for each edge.

Emitted couples ($<$Tag,v$>$,1) are then combined to generate local frequencies (incoming and outgoing degrees) for each source and destination vertex in E_i^{map}. These combined records are then partitioned using a user "defined partitioning function" by hashing only key part v and not the whole mapper tagged key $<$Tag,v$>$. The result of combine phase is then sent to reducers of destination in the shuffle phase of the following reduce step. This step is performed to compute the number of local incoming and outgoing edges to each vertex. The cost of this step is at most :

$$Time(a.1.2) = O\left(\max_{i=1}^{NB\_mappers} c_{comm} * |V_i^{map}| + \max_{i=1}^{NB\_mappers} \|V_i^{map}\| * \log\left(\|V_i^{map}\|\right) \right).$$

The term $c_{comm} * |V_i^{map}|$ is time to communicate data from mappers to reducers, whereas the term $\|V_i^{map}\| * \log\left(\|V_i^{map}\|\right)$ is time to sort mapper's emitted records.

The global cost of this step is therefore :

$$Time_{step_{a.1}} = Time(a.1.1) + Time(a.1.2).$$

a.2: Reduce phase to combine shuffled records and to create global histogram for "high degree vertices partitioning templates"

At the end of the shuffle phase, each reducer i will receive a subset, called V_i^{red}, of vertices (and their corresponding incoming and outgoing local degrees) obtained through hashing of distinct values of V_j^{map} held by each mapper j. In this step, received incoming and outgoing local frequencies are then merged to compute the global frequencies (the global incoming and outgoing degrees) for each vertex present in set V. To this end, each reducer i emits, for each **high-degree vertex** "v", a couple ("v", Nb_LeftSlaveVertices, Nb_RightSlaveVertices) where :

- Nb_LeftSlaveVertices is the number of "left" slaves to create in the following phase b.1; these "left" slaves are used to partition incoming edges to a high-degree vertex "v". Nb_LeftSlaveVertices depends only on the number of incoming edges to vertex "v".
- Nb_RightSlaveVertices is the number of "right" slaves (these slaves are created in phase b.1) used to partition outgoing edges of a high-degree vertex "v". Nb_RightSlaveVertices depends only on the number of outgoing edges of "v".

Using this information, each reducer i has local knowledge of how high-degree vertices will be partitioned in the next map phase.

The global cost of this step is at most $Time_{step_{a.2}} = O\left(\max_{i=1}^{NB\_reducers} \| V_i^{red} \| \right)$.

To avoid the effect of high-degree vertices in graph processing, generated "left" and "right" slave vertices are assigned to distinct "workers" in a round-robin manner to balance load during all the stages of graph processing.

To guarantee scalability and perfect balancing of the load among processing nodes, partitioning templates and graph partitioning are carried out jointly by all reducers (and not by a coordinator node). Note that only edges associated to high-degree vertices are split and sent to distinct workers, whereas edges associated with low degrees for both source and destination vertices are emitted without any transformation.

b.1: Map phase to generate partitioned graph

Each mapper reads the global histogram of "high-degree vertices partitioning templates" from DFS and creates a local hash table.

In this step, each mapper i reads its assigned edge splits of input graph from DFS and generates a set of left and right slave vertices for each edge depending on the degrees of source and destination vertices: only edges associated to a high-degree source or destination vertex are transformed, whereas those associated with low degree for both source and destination vertices are emitted as they are without any graph transformation. Note that left and right slave vertices associated to a high-degree vertex are created only once by a designated worker and new edges are created from each master vertex to its left and right slave vertices.

At the end of this step, mappers emit all edges resulting from this graph transformation and mark high-degree vertices for a later removal. The cost of this step is at most :

$$Time(b.1) = O\left(\max_{i=1}^{NB\_mappers} c_{r/w} * |E_i^{map}| + t_h^i * \| \overline{V} \| + 2 * t_s^i * \| E_i^{map} \| \right).$$

The term $c_{r/w} * |E_i^{map}|$ is time to read edges of input graph from DFS on each mapper i, the term $t_h^i * \|\overline{V}\|$ is time to build the hash table holding high-degree vertices and their corresponding values `Nb_LeftSlaveVertices` and `Nb_RightSlaveVertices`, whereas the term $2 * t_s^i * \|E_i^{map}\|$ is time to perform a hash table search for both source and destination vertices of mapper's input edges. We recall that the size of this hash table is, in general, very small compared to the size of input graph.

The global cost, $Time_{GPHDV}$, of `GPHDV` algorithm is therefore the sum of above three phases, and GPHDV algorithm has asymptotic optimal complexity when:

$$\|\overline{V}\| \leq \max\left(\overset{NB\_mappers}{\underset{i=1}{\max}} \|\|E_i^{map}\| * \log(\|E_i^{map}\|), \overset{NBreducers}{\underset{i=1}{\max}} \|E_i^{red}\| \right); \qquad (8.1)$$

this is due to the fact that all other terms in $Time_{GPHDV}$ are, at most, of the same order of $\|E_i^{map}\|$. Inequality (8.1) holds, in general, since $\overline{V}$ contains only high-degree vertices and the number of these high-degree vertices is very small compared to the number of input graph vertices.

8.4 Experiments

In order to evaluate the performance of our high-degree vertex partitioning approach, we compared the execution of the Single Source Shortest Paths (SSSP) problem on both partitioned and unpartitioned graph data using Hadoop-1.2.1 and Giraph-1.2.0 frameworks.[3] We ran a large series of experiments where 49 `Virtual Machines (VMs)` were randomly selected from Orléans University cluster using OpenNebula software for VMs administration. Each virtual machine has the following characteristics: 1 Intel(R) Xeon@2.53GHz CPU, 2 Cores, 6 GB of Memory, and 80 GB of Disk. Setting up a Hadoop cluster consisted of deploying each centralized entity (NameNode and JobTracker) on a dedicated `Virtual Machine` and co-deploying DataNodes and TaskTrackers on the rest of VMs. The data replication parameter was fixed to three in the HDFS configuration file.

To study the effect of data skew on SSSP performance, we use synthetic graphs following a power low distribution. To this end, generated graphs have been chosen to follow a Zipf distribution [22] as it is the case in most database tests: Zipf factor has been varied from 0 (for a uniform data distribution) to 2.8 (for a highly skewed data). Note that natural graphs follow a power low of $\sim$2 which corresponds to a highly skewed data [9, 16]. In our experiments, each input graph has a fixed size of 200M vertices and 1B edges (corresponding to about $\sim$25 GB for each graph data).

We noticed, in all the tests and also those presented in both figures Figs. 8.4 and 8.5 that using GPHDV graph partitioning algorithm, SSSP problem processing is

---

[3]Hadoop and Giraph frameworks are, respectively, implementations of MapReduce and Pregel developed by "The Apache Software Foundation".

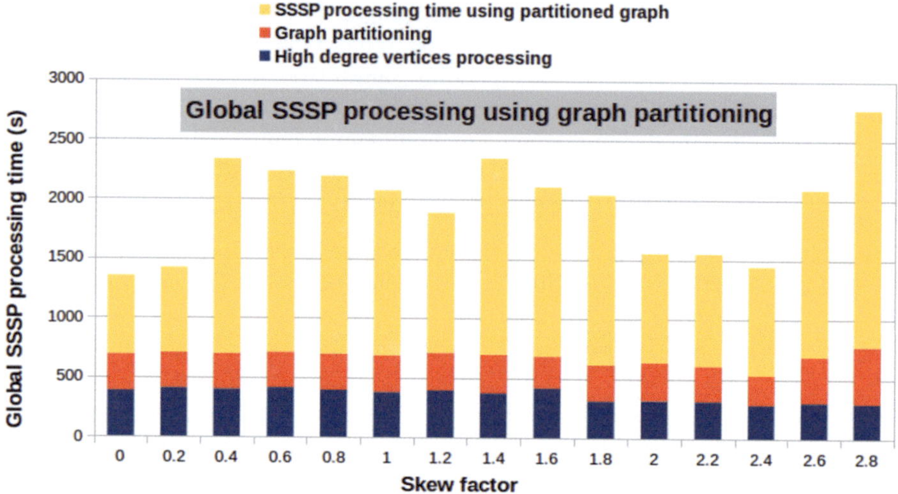

Fig. 8.4 Graph skew effects on SSSP processing time using our graph partitioning approach

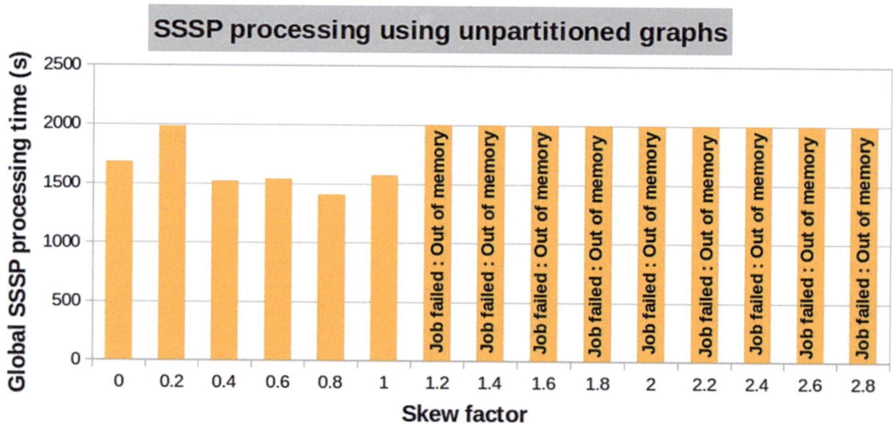

Fig. 8.5 Graph skew effects on SSSP processing time using unpartitioned graphs

insensitive to data skew, whereas the same execution using nonpartitioned graphs
fails due to lack of memory for skew factors varying from 1.2 to 2.8 (see Fig. 8.5).
Moreover, in Fig. 8.4, we can see that GPHDV processing time (which includes
both "high-degree vertices processing" and "graph partitioning") remains very small
compared to Giraph SSSP execution time and that the overhead related to GPHDV
preprocessing is very small compared to the gain in performance related to the
use of partitioned graphs. Note this partitioning is performed only once for each
graph. This performance is due to the fact that, in partitioned graphs, computation
and communication are much more balanced compared to Giraph execution using
nonpartitioned graphs. This shows that GPHDV preprocessing makes Giraph

(and therefore all Pregel-like systems) more scalable and insensitive to the effects of high-degree vertices in large-scale graph processing due to the fact that each vertex has a limited number of incoming/outgoing edges since the number of the neighbors of each vertex never exceed a user-defined threshold t_0 (this threshold is used to set the appropriate number of "left" and "right" slaves of each high-degree vertex) using our partitioning approach. This also avoids communication bottleneck owing to the bounded number of messages sent/received by each vertex, in each graph processing step.

8.5 Related Work

Apache Giraph: [8] A distributed framework for processing iterative algorithms on large-scale graph. Giraph is an open implementation of Pregel [15] inspired by the bulk synchronous parallel (BSP) model [20] and based on a vertex-centric programming model. In Giraph, vertices are divided into partitions assigned to distinct workers. The default partitioning method is based on a hash function (or range partitioning) applied on vertices IDs. In addition, a partitioning method implemented by the developer can be used. Vertices can communicate by sending messages, and periodic checkpoints are executed for fault tolerance. Giraph has extended Pregel API by adding master compute function, out-of-core capabilities that allows to treat main memory limitation by splitting graph partitions and messages to local disks, shared aggregators, edge-oriented input, and so on. This implementation of Pregel is very efficient in big graph processing. However its performance degrades in the presence of high-degree vertices. The performance of Pregel-like systems also degrades in the presence of load imbalance among workers in graph processing supersteps.

GraphLab: [14] An asynchronous parallel framework based on a distributed shared memory architecture. In GraphLab, each vertex program can directly access and accumulate data from its adjacent edges and vertices to update its state. The updated vertex or edge data are automatically visible to the adjacent edges. In this approach, using a shared memory may limit the scalability and the efficiency of big graph processing. Recently, several studies reported that asynchronous graph processing is generally slower than its synchronous mode (used in all Pregel-like systems) due to the high locking/unlocking overhead [21].

PowerGraph: [9] A distributed framework supports both Pregel's bulk synchronous and GraphLab's asynchronous models of computation. PowerGraph relies on Gather, Apply, and Scatter (GAS) programming model to implement the vertex program. In Giraph and GraphLab, the graph is partitioned using edge-cut approach based on hash partitioning, in order to balance the load of different processors. However, hash partitioning is not efficient in the case of natural graph due to the skewed power-law distribution where only few vertices are connected to high number of vertices (e.g., celebrities on social media), while most vertices have few neighbors. To address this problem, PowerGraph follows vertex-cut approach, where high-degree vertices are split over several machines thus balancing the load of machines and reducing communication costs. As mentioned earlier for GraphLab, PowerGraph may also suffer from scalability in big graph processing in large-scale systems.

GPS: [17] An open-source distributed big graph processing framework. It extends Pregel's API with a new *master.compute()* function that allows to develop global computation in an easier way. In addition, GPS follows a dynamic graph repartition approach that allows to reassign vertices during job execution to other machines in order to reduce communication costs. It also has an optimization technique called large adjacency list partitioning (LALP) that allows to partition neighbors of high-degree vertices over many machines. The idea of vertex partitioning in GPS is similar to vertex mirroring and the cost of this partitioning may be very high.

8.6   Conclusion and Future Work

In this article, we have introduced an efficient and scalable MapReduce graph partitioning algorithm based on a master/slaves approach called GPHDV. This graph partitioning algorithm is used (as a preprocessing phase for big graph algorithms) to make Pregel-like systems scalable and insensitive to the problem of high-degree vertices in big graph processing. In this algorithm, incoming and/or outgoing edges associated with each high-degree vertex are partitioned over a set of slave vertices, and these slave vertices are effected to distinct workers in a round-robin manner to balance computation and communication. The performance results proved that this partitioning approach solves efficiently the problem of load imbalance among "workers" where existing approaches fail to handle the communication and processing imbalances due to the presence of high-degree vertices. It also solves the limitations of existing approaches to handle large graph datasets whenever data associated with high-degree vertices cannot fit in worker's local memory. We recall that partitioning in GPHDV is performed by all the mappers and not by a coordinator processor which guarantees the scalability of this algorithm. This partitioning approach also guarantees perfect balancing properties during all the stages of big graph processing.

Future work will be devoted to big data/graph mining using this partitioning and/or similar techniques based on the use of randomized keys redistribution introduced for join processing in large-scale systems [10, 11] to avoid the effects of load imbalance among processing nodes while guaranteeing the scalability of the proposed solutions.

Acknowledgements This work is partly supported by the GIRAFON project funded by *Centre-Val de Loire* region (France).

References

1. Abou-Rjeili, A., & Karypis, G. (2006). Multilevel algorithms for partitioning power-law graphs. In *Proceedings of the 20th International Conference on Parallel and Distributed Processing*, IPDPS'06 (pp. 124–124). Washington, DC: IEEE Computer Society.
2. Akoglu, L., Tong, H., & Koutra, D. (2014). Graph-based anomaly detection and description: A survey. *CoRR*. abs/1404.4679.
3. Apache Hadoop. http://hadoop.apache.org/core/.
4. Aridhi, S. & Nguifo, E. M. (2016). Big graph mining: Frameworks and techniques. *CoRR*. abs/1602.03072.
5. Atastina, I., Sitohang, B., Saptawati, G. A. P., & Moertini, V. S. (2017). A review of big graph mining research. *IOP Conference Series: Materials Science and Engineering, 180*(1), 012065.
6. Dean, J., & Ghemawat, S. (2004). Mapreduce: Simplified data processing on large clusters. In *OSDI'04: Sixth Symposium on Operating System Design and Implementation*, San Francisco.
7. Garey, M. R., Johnson, D. S., & L. Stockmeyer. (1974). Some simplified np-complete problems. In *Proceedings of the Sixth Annual ACM Symposium on Theory of Computing*, STOC'74 (pp. 47–63). New York: ACM.
8. http://giraph.apache.org/.
9. Gonzalez, J. E., Low, Y., Gu, H., Bickson, D., & Guestrin, C. (2012). PowerGraph: Distributed graph-parallel computation on natural graphs. In *Proceedings of the USENIX Conference on Operating Systems Design and Implementation*, Berkeley (pp. 17–30)
10. Hassan, M. A. H., & Bamha, M. (2015). Towards scalability and data skew handling in group by-joins using mapreduce model. In *Proceedings of the International Conference on Computational Science, ICCS 2015*, Reykjavík, Iceland, 1–3 June 2014 (pp. 70–79).
11. Hassan, M. A. H., Bamha, M., & Loulergue, F. (2014). Handling data-skew effects in join operations using mapreduce. In *Proceedings of the International Conference on Computational Science, ICCS 2014*, Cairns, 10–12 June 2014 (pp. 145–158).
12. Kang, U., & Faloutsos, C. (2012). Big graph mining: Algorithms and discoveries. *SIGKDD Explorations, 14*(2), 29–36.
13. Leskovec, J., Lang, K. J., Dasgupta, A., & Mahoney, M. W. (2008). Community structure in large networks: Natural cluster sizes and the absence of large well-defined clusters. *CoRR*. abs/0810.1355.
14. Low, Y., Bickson, D., Gonzalez, J., Guestrin, C., Kyrola, A., & Hellerstein, J. M. (2012). Distributed graphlab: A framework for machine learning and data mining in the cloud. *Proceedings of the VLDB Endowment, 5*(8), 716–727.
15. Malewicz, G., Austern, M. H., Bik, A. J. C., Dehnert, J. C., Horn, I., Leiser, N., & Czajkowski, G. (2010). Pregel: A system for large-scale graph processing. In *Proceedings of ACM SIGMOD International Conference on Management of Data*, New York.
16. Newman, M. E. J. (2004). Power Laws, Pareto Distributions and Zipf's Law. *Contemporary Physics, 46*(5), (323–351).
17. Salihoglu, S., & Widom, J. (2013). GPS: A graph processing system. In *Proceedings of the 25th International Conference on Scientific and Statistical Database Management*, SSDBM (pp. 22:1–22:12). New York: ACM.
18. Skhiri, S., & Jouili, S. (2013). *Large graph mining: Recent developments, challenges and potential solutions* (pp. 103–124). Berlin/Heidelberg: Springer.
19. Tang, L., & Liu, H. (2010). *Graph mining applications to social network analysis* (pp. 487–513). Boston: Springer.
20. Valiant, L. G. (1990). A bridging model for parallel computation. *Communications of the ACM, 33*(8), 103–111.
21. Yan, D., Cheng, J., Lu, Y., & Ng, W. (2015). Effective techniques for message reduction and load balancing in distributed graph computation. In *Proceedings of the 24th International Conference on World Wide Web*, WWW'15 (pp. 1307–1317). Republic and Canton of Geneva, Switzerland. International World Wide Web Conferences Steering Committee.
22. Zipf, G. K. (1949). *Human behavior and the principle of least effort: An introduction to human ecology*. Cambridge:Adisson-Wesley.

Chapter 9
Nature-Inspired Radar Charts as an Innovative Big Data Analysis Tool

J. Artur Serrano, Hamzeh Awad, and Ronny Broekx

9.1   Introduction

Radar charts have been widely used as representations of data varying over a period of time, for example in the scope of climate research [1]. They have also been used for the presentation of healthcare [2] and well-being [3] data. A radar chart visually compares multiple types of data on multiple dimensions. Radar charts are useful for the comparison of points of two or more different data sets. An example is shown in Fig. 9.1.

In our approach, we express the data using a close-to-nature visual representation. The model takes its inspiration in nature. In the model, petals represent the various variables, and both colour and size are linked to properties of the data (Fig. 9.2).

Choice of colours and brightness allow a quick inspection of the data. The properties of the flower are linked to the properties of the health data in a way that when values rise above certain thresholds, the flower looks less healthy as a metaphor to data representing an undesirable health condition. The flower petals in

J. Artur Serrano (✉)
Department of Neuromedicine and Movement Science, Faculty of Medicine and Health Sciences, NTNU/Norwegian University of Science and Technology, Trondheim, Norway

Norwegian Centre for eHealth Research, University Hospital of North Norway, Tromsø, Norway

H. Awad
Health Science Department, Khawarizmi International College (KIC), Abu Dhabi, UAE

Department of Applied Science, College of Arts and Sciences, Public Health Program, Abu Dhabi University, Abu Dhabi, UAE

R. Broekx
Innovation Department, ePoint, Hamont, Belgium

© Springer International Publishing AG, part of Springer Nature 2018

M. M. Alani et al. (eds.), *Applications of Big Data Analytics*,
https://doi.org/10.1007/978-3-319-76472-6_9

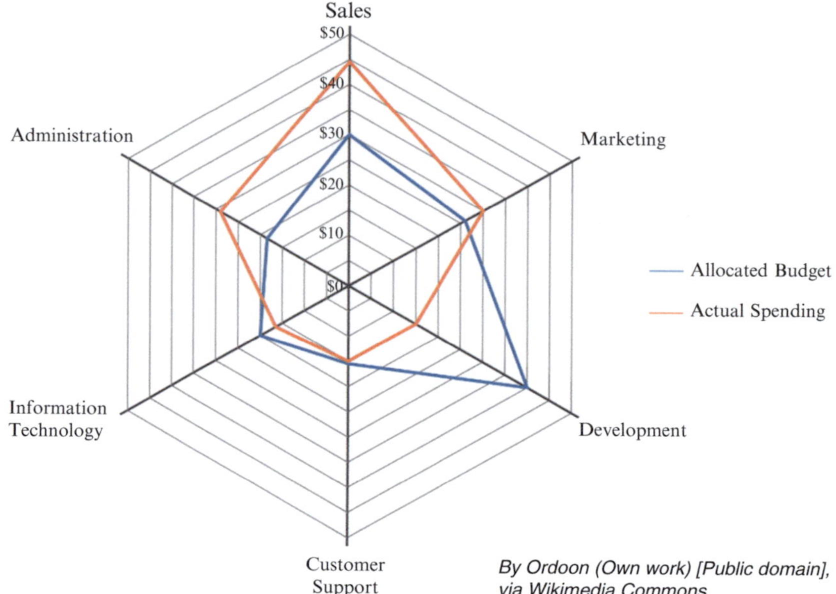

By Ordoon (Own work) [Public domain], via Wikimedia Commons

Fig. 9.1 Example of a radar chart

Fig. 9.2 Sunflower

ideal circumstances look bright and with even colour. This way a patient can look at his/her data without the knowledge of the particular thresholds.

If this system will work for a patient, it will also work for visualization of Big Data for researchers, for example, for the analysis of population health trends. In this case, the individual flower becomes a flower field representing individuals or groups of individuals. Zooming capabilities would allow for detailed analysis of particular sets of flowers. It would be also interesting to geographically compare data amongst different populations, which would be represented as different flower fields. In addition, interaction between populations could be identified, for example, looking at interactions in the health condition of neighbouring flower fields.

The analogy could be taken further and look at long time frames and see which fields are blossoming and in which parts of the world population health is declining.

This approach could be useful to give access to both the general population and researches to interpret Big Data without deep knowledge in specific fields of healthcare. The influence of external factors, such as diet, medication, ageing, etc., in the health condition could also be modelled and visualized with this approach.

9.2   Using a Common Visualization Metaphor

Figure 9.3 shows a representation of a flower in a way that we might find familiar as such images are common in high school books and other educational materials. In this way, this type of representation does not need further explanation, as we all share the same knowledge background for interpreting the data presented.

If we take a flower's details, we can use the various components to represent health data (e.g. the colour of petals to map blood pressure values, glucose levels or cholesterol).

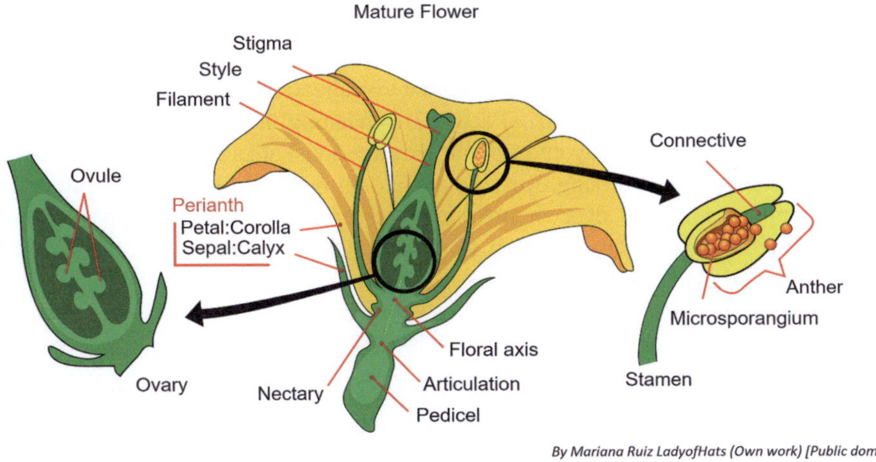

Fig. 9.3 A typical presentation used in educational materials

Variability in flower stems' heights can also be used to indicate, for instance, deviation to a mean value. This would easily provide a group pattern identification by direct visual observation.

In the overview of a particular area, we see which flower meadows do well or do worse. This behaviour can be caused by environmental factors such as nutrition, air quality, working conditions, etc. Epidemiological data and its correlations may this way be depicted so that non-specialists in the health field can easily analyse it.

If we now look at a flower individually, we are then looking not at demographic data but at individual data. This could eventually be used by individuals to track their own health data with recognizable computer-generated imagery. Moreover, such visualizations, due to their reduced size, are suitable to be deployed on a smartwatch.

Since our approach does not rely on a specialized knowledge on the graphical representation, it makes the data widely available over a larger group of professionals, thus promoting multidisciplinary team (MDT) decision-making. This may also facilitate MDT processes with focus on patient involvement.

9.3   Challenges and Problems in Big Data Visualization

As the available amount of data grows, traditional visualization tools will tend to reach their limits as these data are evolving continuously [4]. Data visualization applied to Big Data analytics plays important role in decision-making. But as pointed out by Gorodov and Gubarev [5], it is quite a challenge to visualize such a mammoth amount of data both in real-time and in static form. An additional challenge is the format in which Big Data is typically generated. Most of traditional visualization tools are optimized for structured data, generally stored in large databases. However the format presented by Big Data is usually semi-structured and unstructured. The use of metadata in the visualization technique can only be taken contextually or by tagging when this is present. Limitation of resolution will also play a crucial role in the decisions made on the visualizations used: if we reduce the dimensions of the data set to be visualized, some meaningful patterns may be lost; if we use too much data, we may end up with cluttered visualizations due to the high density of elements. Visualizing every possible data point can lead to overplotting and overlapping and may overwhelm user's perceptual and cognitive capacities [6]. Most of the current visualization tools have low performance in scalability, functionality and response time [7].

Below are some other important Big Data visualization problems [5] and answers provided by nature-inspired radar charts:

- Visual noise: Most of the objects in data set are too relative to each other. It becomes very difficult to separate them. Using our approach, visual noise is avoided by design. In fact, when visualizing a flower field from a distance, although individual flower patterns are lost (leaf size, stem height, etc.), large group patterns emerge. The concept of zooming in the data corresponds in our approach to dive in the flower fields and, in that way, focusing on a smaller group of plants. This represents an intuitive way of pattern analysis and cluster identification.

- Information loss: To increase the response time, we can reduce data set visibility, but this leads to information loss. Again, we avoid this problem by reducing resolution when instantiating from the flower field and increasing resolution when analysing reduced groups of flowers maintain approximately constant the pixel count in the visible field.
- Large image perception: Even after achieving desired mechanical output, we are limited by our physical perception. A visualization transformation from a three-dimensional plant representing an elevated radar chart into a two-dimensional map of pattern coloured fields gives an intuitive output adapted to the natural human visual perception.

9.4 A View to the Future: Adopting Artificial Intelligence and Virtual Reality

Big Data research has mostly been focusing on the quantitative aspects of the information and how these can be conveyed in a comprehensible and useful way. Big Data is extremely helpful with gathering quantitative information about new trends, behaviours and preferences, so it is no wonder companies invest a lot of time and money sifting through and analysing massive sets of data. The development of health related sensors for personal use, together with the introduction and rapid evolution of the Internet of things, is transforming the vision of a Quantified Self into a possible reality [8]. However, what big data fails to do is explain why we do what we do. The concept of "thick" data fills this gap [9]. Thick data is qualitative information that provides insights into the everyday emotional lives of consumers. It goes beyond big data to explain why consumers have certain preferences, the reasons they behave the way they do, why certain trends stick and so on.

The same with our nature-inspired radar charts, they depict large amounts of data, but by analysing high-level patterns, new connections and correlations can be identified which were not recognized before. This approach makes possible the analysis of petabytes of clinical data without deep insight in the medical field. It seems clear that patterns are easier to detect when the data is not presented in complex models.

Artificial intelligence (AI) algorithms could be developed for the intensive analysis of data correlations. The AI system could be able to unveil correlations not yet discovered. AI could detect the correlations and present them as nature-based images, going from semi-structured data to flower field visualization. AI algorithms would be able to collect the necessary data, find the useful correlations and present the visualization as the answers to questions of researchers and clinicians.

In the future, virtual reality (VR) systems could be used to navigate through these fields generated by Big Data technologies. In the same way a pilot in a plane, flying over a field of crops, is able to assert the correct development of the plants; the same concept would be used by a person with a VR device to virtually fly over the crops of health data fields and assert about the situation of population health status.

Recent advances in VR make possible to give each field of data its own scent [10]. So the sense of smell may be used to indicate possible data trends. This will guide the data analyst to study further a particular crop field.

Existing visualization methods for Big Data can face scalability problems given the natural cognitive limits of humans in dealing with complex data representations. The fast expansion of produced data may result in challenges extracting information and gaining knowledge from it [10]. Adopting a well-known representation such as fields of flowers or crops may transmit a feeling of control, given the familiarity of the context. VR applications to Big Data give the possibility of travelling through the natural representations, as one indeed would be in a field of flowers. Commercial platforms are emerging, equally from large players, such as Amazon Web Services [11], Microsoft Power BI [12], Google Cloud Platform [13], and from independent private companies such as Wrangler produced by Trifacta [14].

VR allows for an evolution from planar representations to tridimensional or volumetric visualizations. As clinical data expands from that obtained from current medical tests as imaging and laboratory analysis into continuous sensor-based monitoring in body area networks or remote sensing of vital signs, the amount of generated data is increasing by several orders of magnitude. Applying VR can help overcome human cognitive limitations in dealing with such amounts of data by providing a representation in a form most suitable for human perception.

Spatial analysis or spatial statistics techniques, used to study entities by their topological, geometric or geographic properties, can be applied to clinical data through VR technology. The reduction of the cost of such technologies, allied to the rapid increase in the quality of their technical specifications, such as resolution and latency, is making VR a primary candidate for the exploration of Big Data. The market study report performed by Mordor Intelligence on Global Data Visualization shows that the expected compound annual growth rate (CAGR) is nearly 10% until 2022, representing a growth in value from USD 4 billion in 2017 to USD 7 billion in just 5 years.

The VR technology not only facilitates the understanding of Big Data but also the interaction with it. The several ways of interaction include both the well-known *scaling* and the *interactive filtering*. Scaling [15] allows the user to zoom in and zoom out into the data, creating real-time changes in the representation, for example, from a coloured area (representing a garden), shown as a simple aggregation of pixels, to a complex structure (individual flower) in a form of a structured radar chart. With interactive filtering [16] the user may identify relevant data subsets in real time.

In conclusion, our proposed multivariate and dynamic data representation model based on natural visualizations of flower fields, which themselves are composed of individual plant specimens inspired by radar charts, can offer a suitable platform for the implementation of VR solutions. This strategy may facilitate the human exploration of large amounts of healthcare data generated by the present Big Data applications and can be an important instrument prepared to tackle the challenges ahead in this research area.

References

1. Climate spirals. Climate Lab Book – Open climate science. http://www.climate-lab-book.ac.uk/files/2016/05/spiral_optimized.gif. Last accessed 27 Mar 2017.
2. Saary, M. J. (2008, April). Radar plots: A useful way for presenting multivariate health care data. *Journal of Clinical Epidemiology, 61*(4):311–317. ISSN 0895–4356, https://doi.org/10.1016/j.jclinepi.2007.04.021. http://www.sciencedirect.com/science/article/pii/S0895435607003320
3. Li, X., Hong, W., Wang, J., Song, J., & Kang, J. (2006). *Research on the radar chart theory applied to the indoor environmental comfort level evaluation.* 6th World Congress on Intelligent Control and Automation, Dalian, pp. 5214–5217. https://doi.org/10.1109/WCICA.2006.1713386.
4. Ali, S. M., Gupta, N., K. Nayak, G., & Lenka, R. K. (2016). *Big Data visualization: Tools and challenges.* 2nd International Conference on Contemporary Computing and Informatics (IC3I), Noida, pp. 656–660. https://doi.org/10.1109/IC3I.2016.7918044.
5. Gorodov, E. Y., & Gubarev, V. V. (2013). Analytical review of data visualization methods in application to Big Data. *Journal of Electrical and Computer Engineering*, Article ID 969458, pp. 1–7.
6. Tavel, P. (2007). *Modeling and simulation design.* Natick: AK Peters Ltd.
7. Lidong, W., Guanghui, W., & Alexander, C. A. (2015). Big Data and visualization: Methods, challenges and technology progress. *Digital Technologies, 1*(1), 33–38. https://doi.org/10.12691/dt-1-1-7.
8. Gurrin, C., Smeaton, A. F., & Doherty, A. R. (2014). LifeLogging: Personal Big Data. *Foundations and Trends in Information Retrieval, 8*(1), 1–125. https://doi.org/10.1561/1500000033.
9. Intel IT Center. (2013, March). *Big Data visualization: Turning Big Data into big insights.* White Paper, pp.1–14.
10. FeelReal. http://feelreal.com/#. Last accessed 14 Dec 2017.
11. Cook, J. *The power of thick data.* Big Fish Communications. http://bigfishpr.com/the-power-of-thick-data/. Last accessed 02 Sept 2017.
12. Shull, F. (2013, July/August). *Getting an intuition for Big Data.* IEEE Software, pp. 1–5.
13. Kim, Y., Ji, Y.-K., & Park, S. (2014). Social network visualization method using inherence relationship of user based on cloud. *International Journal of Multimedia and Ubiquitous Engineering, 9*(4), 13–20.
14. Olshannikova, E., Ometov, A., Koucheryavy, Y., & Olsson, T. (2015). Visualizing Big Data with augmented and virtual reality: challenges and research agenda. *Journal of Big Data, 2*(22).
15. Amazon Web services. https://aws.amazon.com. Last accessed 02 Sept 2017.
16. Power BI. https://powerbi.microsoft.com. Last accessed 02 Sept 2017.

Chapter 10
Search of Similar Programs Using Code Metrics and Big Data-Based Assessment of Software Reliability

Svitlana Yaremchuck, Vyacheslav Kharchenko, and Anatoliy Gorbenko

10.1 Introduction

IT industry and software engineering as a kernel of information technology is one of the key branches for global economics. Cloud computing and the Internet of things, machine learning and artificial intellect, software-defined and hyper-converged infrastructures, and so on impact on the business and service quality more and more. On the other hand, human safety and security more and more depend on reliability of software and software-based systems. Software faults and software-based system failures cause environmental disasters, loss of billions dollars, and damage to health and life of people.

To assure the required level of software reliability, the high accuracy of reliability assessment and indicators evaluation must be provided. Assurance and approving of required software reliability is one of the challengeable problems for software engineering. There are two paradoxes in software reliability and its quantitative assessment:

S. Yaremchuck (✉)
Department of General Scientific Disciplines, Danube Institute of National University "Odessa Maritime Academy", Odessa, Ukraine
e-mail: s.yaremchuck@csn.khai.edu

V. Kharchenko
Computer Networks and Systems Department, National Aerospace University "KhAI", Kharkiv, Ukraine
e-mail: v.kharchenko@csn.khai.edu

A. Gorbenko
School of Computing, Creative Technologies & Engineering, Leeds Beckett University, Leeds, UK
e-mail: A.Gorbenko@leedsbeckett.ac.uk

© Springer International Publishing AG, part of Springer Nature 2018
M. M. Alani et al. (eds.), *Applications of Big Data Analytics*,
https://doi.org/10.1007/978-3-319-76472-6_10

- Dynamical development of software technologies and increasing number of implemented software products are followed by increasing of reliability model number (as software reliability growth models (probabilistic SRGMs), deterministic, and other types of models) to evaluate quantitative metrics and indexes. However, engineering confidence in these models and intensity of their application are decreased. This paradox can be called as "models for models" (M-paradox).
- Data capacity regarding software reliability (test cases, software faults, etc.) is increased, but such data are usually used to prove importance of reliability improving or to specify input parameters for evaluation. That is, huge information is not systematically processed and applied to predict the reliability level. Such situation is like an astronomer who searches new stars without studying books about the star sky (A-paradox).

M-paradox is caused by three main reasons:

- The insufficient determinacy in customer requirements to the reliability level.
- The academism and using complexity of software reliability models, first of all, are probability models, because these models require the determination of input parameters values.
- The insufficient level of technological effectiveness and imbedding of procedures and tools of reliability assessment in development processes, verification, and validation of software.

A-paradox is stipulated by shortage of modern technologies for searching, systemizing, and processing data about software faults and hardware failures (electronic components, chips, and so on).

The data on faults and reliability indicators become available in the process of software testing and using. These data are stored in various storages and are provided by various services. We consider that these data can be used in the course of creation of new systems to increase in their reliability. We offer three possible scenarios for reliability software estimation: the scenario S1, to develop, verify, and use new reliability model for new software (only for this project!); the scenario S2, to select one of the known models, to verify this model, and to use it for this project (and for other projects as well); and the scenario S3, to find the reliability data for similar software systems according to testing and operation results, to process these data, and to calculate reliability indexes of the new system (and for other projects as well). The scenario S3 is directed to estimation of reliability software under development on the basis of reliability data analysis of software which was developed, tested, and used earlier and has similar characteristics. Within this article the scenario S3 is researched.

Application of the big data analysis techniques is relevant and economically motivated to assess and improve system and software reliability. The objective of this research is to adapt the big data analysis methods for tasks of estimation and prediction and increase of software reliability. We offer the following: firstly, a technique to search of the so-called similar programs on the basis of similarity of source code metrics to use data on their reliability; secondly, a set of indexes for reliability estimation of similar programs code; and thirdly, recommendations to use the indexes for software reliability prediction.

Structure of the chapter is the following. In Sect. 10.2 we describe the state of art in the field of software reliability and suggest adapting two methods of big data analysis for the tasks of software reliability estimation. In Sect. 10.3 we offer the main approach to software reliability prediction. The main approach means that the program systems, similar in the properties, have similar reliability. In Sect. 10.4 we describe the concept of software systems similarity, metrics, and procedure and search results of similar program. In Sect. 10.5 we adapt Map and Reduce-based procedure for search, preliminary processing, reducing of set data size, and reliability analysis of similar programs. In Sect. 10.6 we offer reliability indexes and show their analysis, visualization, and interpreting. In Sects. 10.7 and 10.8, we discuss research results and formulate key questions of further investigation correspondingly.

10.2   State of Art

Researches developed a number of models, methods, and software application to evaluate and manage the software reliability [1–8]. Many research works of the recent years are directed to the increasing of software reliability on scenario S1 base. The work [1] has been devoted to the basic theory of the of software systems dynamics and established the theoretical basis for the reliability assessment and proposed a new universal model for such assessment. The proposed model requires experimentally obtained data of faults revealing time in the course of testing. The more data become available, the more precise are reliability evaluating indicators, the shorter is the testing period, and the less are resources required to achieve the required reliability level. Besides, software company specialists need to calculate model parameters.

The works [2, 3] describe software reliability model with complexity index based on the nonhomogeneous Poisson process (NHPP). The authors offer software application for software failures prediction on the basis of artificial neural networks. However, reliability prediction by means of neural networks makes the developers use the software known not so well and to study network characteristics. The complexity prevents this method from implementation into routine engineering practice. The work [4] describes the new analytic model, establishing the relation between faults quantity in the initial code and complexity indexes based on the code metrics. Application and statistical analysis of the proposed method showed that discrepancy between actual faults quantity and the estimated one amounted to 11%.

The work [5] researches methods of fault localization in software modules. As a result, 9% discrepancy between the obtained indicators and the actual quantity of faults has been found. However, the problems associated with choice and ranging of the faults-prone software modules have not been solved yet. The works [6, 7] are directed to execution of scenario S2. In the work [6], many different existing models and methods of evaluation of the quality and software reliability are classified. In this work different models (about 50) are described, but recommendations for the right choice of model are not developed enough. The authors of the work [7] suggested a method based on matrix allowing of the known reliability models. The models are difficult in use and demand from software companies specialists of specific knowledge and skills.

The article [8] describes the faults analysis approach that the authors suggest evaluating the software development process with respect to the quality of produced software and its relation with the required effort. For search of up-to-date information in use of the scenario S3, we carefully studied more than ten sources on big data analysis.

In the course of the review and the analysis of [9–24], we found a lot of information on the use of big data analysis for increasing software reliability. In the research [9], the authors describe the models, methods, technologies and tools for processing, the analysis, visualization, and use of big data. However in this detailed analytical report, not is a word written about the use of big data in software development process. The authors of the research [10] mark barriers and feature implementation of big data. The authors define the following barriers sequence: shortage of specialists, lack of time taken to analyze, high cost of big data storage and manipulation, and high cost of big data analysis. The authors of the research [11] show the basic schematic data management approach on Fig. 10.1. In our opinion, this general approach corresponds completely to processes of collection and big data analysis of software reliability.

The input filter eliminates incorrect and incomplete data. The use of metadata is necessary for establishment of relevance, the sizes, structure, sources, and data

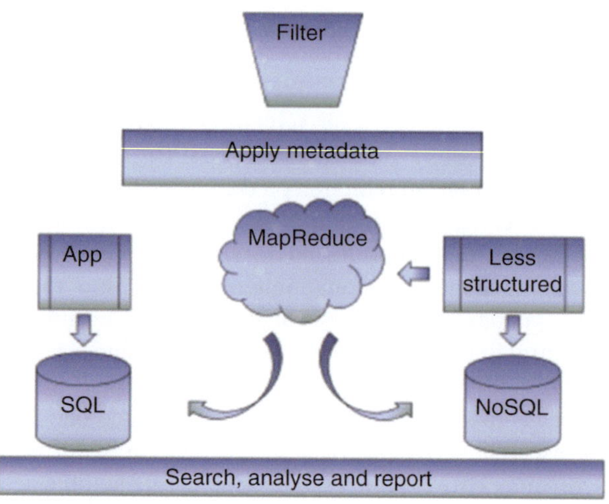

Fig. 10.1 The basic schematic data management approach

components. Metadata define an opportunity, width, and depth of the analysis. The Map step executes data distribution for processing nodes, search of a similar system, and preliminary processing of its data. The Reduce step makes the convolution and analysis of similar program data. The structured data (system properties estimates on metrics, a faults number, time series of faults) boot in local or cloudy corporate SQL storages. The semi-structured data (requirements, the source code) and unstructured data (pictures, diagrams, charts, audio and video records) boot in NoSQL storages. The search of data sets necessary for the analysis is executed in these storages. The final analytical report contains data visualization and numerical indexes of reliability. Such report is used by quality analysis managers and verification specialists for software reliability increase. However, the process of similar system search and the analysis of its data require the appropriate software tools.

The authors of the research [12] notify that deploying a data lake lets companies absorb information from a wide variety of sources in a single repository. Extracting, transforming, and loading (ETL) are the most prevalent data integration processes used by IT and analytics teams. Common characteristics of modern ETL tools are support for integrating data stored in both on-premises and cloud systems, including hybrid cloud environments that mix the two types of platforms. There are dozens of vendors you can consider in your search for the best ETL tools to handle the data integration jobs your organization needs to run. Major vendors selling full integration suites include Oracle, Informatica, IBM, SAP, SAS, Microsoft, and Information Builders. Also available, potentially at a lower cost, are open source ETL and integration platforms from companies such as Talend, Pentaho, and TIBCO Software's Jaspersoft unit.

The author of the research [13] notify that, once data has been loaded from Oracle Big Data Appliance into Oracle Database or Oracle Exadata, end users can use one of the following easy-to-use tools for in-database and the advanced analytics: Oracle R Enterprise, in-database data mining, in-database text mining, in-database graph analysis, in-database spatial, and in-database MapReduce. The authors of the analytical report [14] point out that the security of big data is increasingly as critical for business as it is for governments. The EU Directive 95/46/EC requires data controllers to implement technical and organizational measures to protect personal data against accidental or unlawful destruction, accidental loss, alteration, unauthorized disclosure or access (particularly when the processing involves the transmission of data over a network), and against all other unlawful forms of processing. The authors notify of ten risk-mitigation practices for a big data project. In the report [15], experts discuss different issues: Hadoop users mine big data's business benefits, real-time streaming speeds big data analytics, living with real-time data loss, when spark cloud deployments make sense, and some other problems of big data.

According to the Gartner report [16], during 2017, 60% of big data projects will fail to go beyond piloting and experimentation and will be abandoned. During 2018, 90% of deployed data lakes will be useless as they are overwhelmed with information assets captured about uncertain use cases. The authors of the work "Best Practices for a Successful Big Data Journey" [17] define and describe five phases of a big data journey: ad hoc, opportunistic, repeatable, managed, and optimized phases. The authors notify: as your company evolves through these phases, you will see an exponential increase in value. The authors of the article [18] focused on hardware reliability. Reliability field data have been used for such purposes as generating predictions for warranty costs and optimizing the cost of system operation and maintenance. The work [19] is one of the few researches of software reliability. In order to improve software reliability, the author of this work proposes a novel approach that applies data mining techniques to extract useful information from large software including source code and documents. The paper [20] is one of the few researches in software reliability field with the use of big data. The proposed approach is called Methodology of Technology Oriented Assessment of Software Reliability (TOAStRa). This paper describes the concept, tasks, and some solutions of the TOAStRa methodology.

The search and use of similar entities are widely known and often used big data analytics method. In the work [21], the authors offer a formula for numerical estimation of coefficient of various data sets similarity. The model of the distributed calculations MapReduce is developed by the Google Company for big data analysis [22]. According to MapReduce, the data processing consists of two steps – Map and Reduce. On Map step one of the computers (called master node) obtains input data of the task, divides them into parts, and transfers them to other computers (called worker nodes) for the subsequent processing. On Reduce step convolution and data analysis is made. A function which realizes a convolution algorithm is called "hashing function" and is described in the work [23]. The MapReduce model prevents the partial losses and provides the final result of big data processing [24]. We think that it is expedient to apply the method of the search and the analysis of similar systems data sets to the estimation of software reliability. It is also necessary to adapt the MapReduce model for the prediction of software reliability.

However, the methods described in the articles [1–8] do not take into due account the variety and specific features of software development business processes and require great expenditures for adaptation and implementation. The above factors prevent implementing widely available models, methods, and reliability aids in software engineering routine practice, making such implementation impossible in a number of cases. Therefore, it is especially necessary to provide adaptability, high estimation accuracy, imbedding into business processes of software companies, simplicity, and low costs contributing to easy implementation of the assessment technique.

The review and the analysis of these works allow concluding that big data methods and tools of the analysis are widely used and are promptly developing. However we have found very few information on the use of the big data analysis for reliability increase of program systems. The use of big data analytics for estimation and prediction of program systems reliability is in the initial stage of the development. Insufficient use of big data potential in the course of development and operation of software systems creates serious barriers on the way of ensuring their reliability. In our opinion, one of the essential barriers is difficulties in creation of information field of the company, sufficient for the analysis of reliability. For the analysis of reliability, we suggest using the artifacts of development process and the actual reliability indicators of the systems, developed earlier, which are received after their verification.

Such data do not arrive from network an infinite stream, as at online trade or on social networks. Development and verification of a program system take months. For expansion of the information field of the company, we suggest looking and using (except its own systems) the so-called similar program systems with the known reliability data in public network storages.

However several problematic issues interfere with this search and the analysis. The first question is lack of the formalized concept of systems software similarity, complexity of search criteria formalization, and lack of search software tools. The second question is lack of any catalog of reliability data resources. These data storages are well known for reliability researchers in global data field. They may be, for instance, software reliability data depositories [25] of international conferences, NASA data portal [26], services of code testing and statistical analysis [27], software sport services [28], and other sources. The abovementioned big data storages contain gigabytes of codes and experimental reliability data on multiple systems by different developers. This data may consist of artifacts – requirements, initial code, tests, artifacts evaluations, and processes of their development under various metrics. These data include a number of faults in software modules, temporal series of faults detection. These data enable to evaluate and predict reliability indexes. However, these resources are insufficiently known to program engineering specialists. The next task is adaptation of existing methods and development of new methods and software tools for reliability data analysis. For the solution of these problematic questions, we offer a new approach to predict the program systems reliability with the use of big data analytics.

10.3   Main Approach

The main hypothesis is the following: the program systems, similar in their properties, have similar reliability. Proceeding from it, our approach assumes:

1. The search of program systems (with the known reliability data) which are similar to the new system under development
2. The processing, data analysis of these similar systems and estimation of reliability metrics
3. The use of reliability metrics of similar systems for forecasting of new systems reliability

We consider that for realization of the offered approach it is expedient to adapt a method of the search and the analysis of similar data. For determination of systems similarity, we suggest to estimate a program code of earlier developed systems and new system on the basis of metrics that then compare these estimates. The system with the minimum deviations of estimates will be most similar to the new system.

For implementation of the offered approach, it is expedient to adapt the MapReduce model. To find similar system, it is necessary to analyze a large number of systems. The master server divides processing of these systems on working servers into Map step. Each working server calculates deviations of all metric estimates of the assigned systems and transfers deviations to a master server. On Reduce step the master server analyzes deviations of all systems and selects a similar system with the minimum deviations. The working server with similar system selects metric estimates which is the most informative for estimation of its reliability and transfers them to a master server. The master server displaces estimates of a similar system in uniform assessment of properties, calculates indexes of reliability, and analyzes their dependences on this assessment. The found indicators and dependences allow to predict the reliability of the new system. Thus, the selection and the analysis of data turn information of a similar system into knowledge of reliability of the new system prior to her verification. This knowledge allows planning resources and processes of verification and refactoring, to direct efforts of specialists to faulty modules.

So, we need data of similar systems. The following question is: what is the similarity of program systems?

10.4 Concept of Software Systems Similarity

In our opinion, the concept of the software systems similarity is based on five principles. *The first principle* is a similarity of the source program code. *The second principle* is a functional similarity of systems. This factor assumes comparing web browsers with web browsers, the databases with the databases, and operating systems with operating systems in the course of search of a similar system. *The third principle* is a similar qualification of developers. *The fourth principle* is a similarity (for the different companies – developers) or the invariance (for one company) of the development processes. *The fifth principle* is extent of reuse of a faultless code of own or third-party development as a part of the new system. We believe that the offered list of the similarity principles can be added in the process of further research.

In the work [20], we considered only the first principle. It was the principle of a source program code similarity. In our opinion, the code similarity means the similarity of its properties – the sizes, structure, complexity, and a programming language through proximity of the corresponding metrics values.

10.4.1 Metrics and Procedure

We used metrical data including the estimates on 20 metrics to estimate the similarity of a program code of various systems. These metrics (e.g., KLOC, RFC,

WMC, LCOM, CBO, CE, CA, NPM, DIT, and NOC) have been offered in the work [29] to obtain assessments of the object-oriented source code. The possibility to assess features of system being developed (new system) and similar system by means of uniform set of metrics guarantees resemblance of programming languages of these systems. The system similar to the new system should be defined as a system having minimum deviations of ratings by nominated metrics. The selection of a similar system can be made via relative deviations of each of appropriate metrics for the new system and the systems taken for comparison. We calculated the total, average, and maximum evaluation for each metric. Further we calculated the deviations according to the following formulas. We calculated the relative deviation of initial code dimensions by the following formula:

$$RD_{\text{size}} = \frac{|\ KLOC_d - KLOC_c\ |}{KLOC_d} \cdot 100\% \tag{10.1}$$

The bottom index "d" corresponds to the rating of the system under development. The bottom index "c" corresponds to the rating of systems taken for comparison. Code modules quantity relative deviation was calculated by the following formula:

$$RD_{\text{mod}} = \frac{|\ MC_d - MC_c\ |}{MC_d} \cdot 100\% \tag{10.2}$$

with MC – number of modules. Summarized rate relative deviation was calculated by the following formula:

$$RD_{\text{sum}} = \frac{|\ \text{Sum}_d - \text{Sum}_c\ |}{\text{Sum}_d} \cdot 100\% \tag{10.3}$$

with Sum – total evaluation on each metric for all modules. Average rate relative deviation was calculated by the following formula:

$$RD_{\text{avg}} = \frac{|\ \text{Avg}_d - \text{Avg}_c\ |}{\text{Avg}_d} \cdot 100\% \tag{10.4}$$

with Avg – average evaluation on each metric for all modules. Maximum rate relative deviation:

$$RD_{\text{max}} = \frac{|\ \text{Max}_d - \text{Max}_c\ |}{\text{Max}_d} \cdot 100\% \tag{10.5}$$

with Max – maximum evaluation on each metric for all modules.

At the next stage, the calculated deviations were grouped into three. The first group indicates the dimensions deviation rate, the second group indicates the structure deviation rates, and the third group indicates the code complexity deviation rates. The average deviation rate was calculated for each group. Deviations within a group are feasible to apply with unequal priority indexes for system similarity assessment. Under certain circumstances, priority indexes for system similarity assessment may be either dimension, or structure, or complexity of the system. The common general average deviation rate for all the rates was also calculated. This value is feasible to apply for indexes with equal significance.

So, the search for the comparative system similar to new system requires to know the code dimensions; the number of modules; the estimated complexity, evaluated by applying the unified set of metrics; calculated metrical rates deviation; and, finally, the selection of system with minimum deviations. The technique of similar programs search described in paper [20] consists of seven steps.

Step 1. To calculate metrical rates for the structure, dimensions, and complexity for the new system

Step 2. To download consistently identical metrical rates and data of faults for other systems from big data storage

Step 3. To transform the downloaded data into appropriate format for processing

Step 4. To calculate internal deviation rates and general average deviation rate

Step 5. To record deviation rates for each system into the resulting report and sort the rates

Step 6. To select a similar system with minimum deviation rates in the report

Step 7. To calculate reliability indicators of the selected system to predict reliability of the new system

10.4.2 Search Results of a Similar Program

Manual processing of such data may take too much time and effort. The specialized software agent for search of similar programs (SASS) could be helpful in the aspect of automation of such a process [20]. SASS performs the following functions:

1. Data transformation into appropriate format for processing (*.db, *.xml, *.xlsx, etc.)
2. Transformed system data import into SASS memory
3. Data processing – calculation of deviation rates within the groups and total average deviation
4. Entering deviations for each system into the resulting report
5. Deviations sorting to ground a similar system choice

SASS creates the resulting report with groups and average deviations for multiple involved systems. After the deviation values are sorted, the system with minimum deviations from the new system is placed at the top of the resulting report. The report enables to make a well-grounded choice of the system with the highest similarity index (with the minimum deviation) to the new system. Experimental data about faults of the chosen similar system may be used to predict the reliability of the new system. The proposed SASS is a program for processing flat (not linked) tables and for calculation of statistic indexes.

The metrical data and the data about faults for 30 systems have been randomly selected and downloaded from the big data storage [25] into local computer disk. We have chosen 21 systems for the analysis after the check of the integrity, completeness, and sufficiency of data. One of these systems has been taken randomly as a reference point. Other 20 systems have been explored for the similarity of their features (structure, dimensions, and complexity) to the reference system. The metrical data of each system included estimates on 20 metrics. We have chosen 10 metrics with the greatest correlation with faults number from 20 metrics. We used the initial code dimension in thousands of lines (KLOC) for the numerical assessment of the system size. We used the total quantity of code modules, DIT, and NOC metrics for the numerical assessment of system structure. We used RFC, WMC, LCOM, CBO, CE, CA, and NPM metrics for the numerical assessment of system complexity. These metrics have been offered in the work [29] to obtain the properties assessments of the object-oriented source code. Programming language similarity of the systems in question has been supported by the unified set of metrics. The agent SASS designed by the authors transformed data from *.txt or *.csv format into *.dbf format. Further calculations of relative deviations for metrical rates have been performed by means of SQL instructions for each system. Group and average deviation rates have been stored in the resulting report, as shown in Table 10.1.

Table 10.1 Metric rates relative deviations of systems compared with the reference system

№ system	Metrical rates deviations, %			
	Structure	Dimensions	Complexity	Average rate
1	0,0	0,0	0,0	0,0
2	**5,1**	**9,0**	**5,4**	**6,5**
3	12,2	20,6	41,8	24,9
4	12,7	51,3	35,7	33,2
5	12,8	19,0	28,7	20,2
6	14,4	42,1	46,7	34,4
7	22,4	87,7	34,9	48,3
8	23,6	57,4	33,6	38,2
9	24,5	82,3	29,2	45,3
10	27,4	56,0	43,9	42,4
11	28,6	40,5	40,5	36,5
12	28,6	51,9	44,5	41,7
13	29,9	68,0	20,9	39,6
14	30,3	70,9	26,0	42,4
15	30,9	83,6	45,1	53,2
16	31,4	78,9	38,6	49,6
17	46,0	24,7	44,3	38,3
18	71,8	52,5	24,5	49,6
19	74,6	59,0	25,4	53,0
20	270,1	815,7	52,5	379,4
21	350,6	660,9	66,9	359,5

Rates in Table 10.1 have been sorted in the increasing order. The reference system has number 1. Naturally, deviation rates in the corresponding line are zero. System number 2 follows directly after it with minimum deviation from the reference system (highlighted line in Table 10.1). The systems with increasing deviation rates are placed downward. The resulting report enabled us to choose system with the highest level of similarity (with the minimum deviation) to the reference system. Data processing with the use of the agent has taken about 2 h.

The case study has allowed obtaining some experimental results. A system has been identified with minimum 5,1%, maximum 9%, and average 6,5% relative deviation of metrical rates among the 20 explored systems. The obtained results confirm the allegation that systems with known reliability indexes similar to the system under development may be found from great quantity of experimental data kept in big data storage to assess and predict its reliability.

Thus, the system with the similar source code has been found. How to prepare the data of this system for the analysis? The response to this question is in the next section.

10.5   Map and Reduce Steps: Preliminary Processing and Reducing of Data Set Size

According to the model Map and Reduce, the data processing consists of two steps – Map and Reduce. On Map step one of the computers (called master node) obtains input data of the task, divides them into parts, and transfers them to other computers (called worker nodes) for the subsequent processing. In our case the entrance data are the metric data of various program systems. Worker nodes process these data. By the results of the data processing, master node chooses the most similar system and obtains her data with worker node.

On Reduce step convolution and data analysis is performed. In our case master node carries out the normalization and the convolution of metric data of similar system. On the basis of the obtained data, master node carries out the estimation of reliability metrics. Further we offer the analysis of reliability metrics and their use for reliability prediction of new system. These activities are described in more detail below.

To find the similar system, it is necessary to analyze a large number of systems. We suggest Map and Reduce procedure [22, 24]. On Map step the master server distributes the processing of these systems to the working servers. The working server calculates the metric estimates deviations of properties for the assigned systems according to the formulas (10.1), (10.2), (10.3), (10.4), and (10.5). Further the working server transfers the obtained deviations to the master server. On Reduce step the master server analyzes the obtained deviations and selects the similar system with the minimum deviations. The working server executes the preliminary processing of similar system data. The preliminary processing of data consists of the selection of such metric estimates which are the most informative for the reliability estimation of the similar system.

The data sets of systems from the source [25] include the numerical metric estimates of the size, structure, and complexity of the source code for each system module on 20 metrics. The data sets include also the faults number (or the faults absence), revealed in each module during the verification. The data structure is described by the following vector of metrics (VM)

$$VM = \left\{ \begin{array}{l} modulename, wmc, dit, noc, cbo, rfc, lcom, \\ ca, ce, npm, lcom3, loc, dam, moa, mfa, \\ cam, ic, cbm, amc, \max cc, avg cc, bug \end{array} \right\}$$

We give an example a vector of metric assessment (VMA) for some module $VMA = \{$module$_name, 35, 2, 0, 16, \quad 103, 317, 3, 16, 28, 0.813, 865, 0.888, 0, 0.689, 0.152, 1, 10, 23.46, 3, 1.057, 5\}$ It is expedient to reduce the volume of the analyzed data. It is necessary to take the most informative indicators from data set and to ignore the other indicators. In our opinion, the most informative indicators are those which are more closely connected with the faults number (bug in VM). We used the coefficient of pair correlation (correlation coefficient, CC) between the metric estimates and the faults number in the modules for the measurement of the communication degree.

We investigated 20 data sets of various program systems from the storage [25] for the clarification of the nature of the correlation communication. During the research we have revealed the positive and the negative CC and the significant (CC > 0.5, e.g., 0.78) and the insignificant (CC < 0.1, for example, 0.02) CC. The negative CC always had the insignificant values. The research of CC has shown that it is expedient to leave the estimates with CC $\approx$ 0.5 or CC > 0.5 in a data set. The estimates with CC < 0.5 should be excluded from the analysis. Such estimates have the weak correlation with the faults. They are insufficiently informative for the analysis of the system reliability. Thus, on a Map step, we have executed the selection of such metric estimates which are the most informative for the reliability analysis. As a result, VMA has been reduced from 20 to 10 metrics.

The metric estimates of similar system differ in the absolute values in tens and hundreds of times. For an example, in one studied system, the maximum metric estimates of modules were DIT, 8; NOC, 29; NPM, 122; WMC, 130; RFC, 391; LOC, 4275; and LCOM, 7399 units. In this case it is difficult to receive the general assessment for each module properties. It is problematic to combine in one chat the dependences between the reliability and the module properties, including the use of a logarithmic scale. Therefore a need arises of normalization of metric estimates. The normalization allows to lead all used numerical estimates to the approximate area of their change and then to transform them to the single combined assessment of the module properties. The most widespread ways are a linear and nonlinear normalization. The comparative analysis of these methods of normalization demonstrated the following. We found out that a nonlinear normalization of estimates with the use of a hyperbolic tangent has a more composite algorithm of calculation; however it does not give any advantages in this context. Therefore we executed the transformation of initial estimates to normalized ones by means of the linear normalization formula

$$X_{ik}^{n} = \frac{X_{ik} - X_{\min i}}{X_{\max i} - X_{\min i}} \tag{10.6}$$

where X_{ik} is the value for i metric assessment for k module in traditional units of measurement and X_{ik}^{n} is the corresponding normalized value. Then all the selected estimates were normalized according to the formula (10.6) and summarized in the single *combined assessment (CA)* of module properties by a formula

$$CA_{\mathrm{mod}} = \sum_{i=1}^{m} X_{ik}^{n} \tag{10.7}$$

where m is quantity of the selected metric estimates.

The higher CA values are, the more complex the structure, methods, and interrelations of the modules are. Further there was a question of the informational content of the CA values for the analysis of reliability. To answer this question, we investigated the CC between the CA values and the faults numbers in modules. We established that the CC was higher than the highest CC for the selected estimates. It allows concluding that the selected metric estimates were reduced to the uniform CA without the decreasing of correlations with faults and without loss of the informational content of CA values for the reliability analysis.

Thus, we received the uniform CA instead of 20 metric estimates for each module of similar system after execution of Map and Reduce steps. We reduced an array of original data by 20 times. It allows considerably lowering the hardware loading and the processing period, which is especially important during the big data processing.

Except CA, for the properties estimation of the separate module, we suggest using the following derivative estimates. They are CA_{avg}, $CA_{\max}$, and CA_{total} for all modules of the system. CA_{total} shows the structure, the complexity, and the size of the system by means of one number. For example, for the studied systems of CA_{total}, the values were 189, 340, 538, 790, 3459, and 3825 units. The minimum CA_{total} value differed from the maximum value by 18 times. It means that the studied systems differed in the properties by 18 times. The offered estimates simplify estimation, visualization, the properties analysis, and the reliability analysis of the system. These estimates can be used for the systems comparison and the justification of their development costs.

Thus, the application of big data analysis technique MapReduce in assessing software reliability on the basis of similar programs consists of the two following steps:

1. Allocating big data processing of various systems on work servers, data processing performance, the choice of similar system, and the selection of her metric data, which is the most informative for estimation of reliability (Map step)
2. Reducing of metric data dimensionality by means of the normalization and the convolution (Reduce step)

The processed data of the similar system are ready for the analysis after the performance of Map and Reduce steps. However there are the following questions. How can we take the concrete reliability metrics and their dependence on properties of system from these data? How can we turn these metrics into knowledge? How can we use this knowledge for the reliability increase of the new system with simultaneous cutting of costs for its verification? Answers to these questions are given in the next section.

10.6 Case Study: The Research of Reliability Metrics for Similar Systems

In the work [20], we received the results of programs similarity estimation on the basis metric estimates of the structure, the size, and the complexity of the source code by mandatory similarity of a programming language. Later we specified and expanded the list of the researched systems from the resource [25]. After specification and addition, the results of similarity estimation for 22 systems have been represented in Table 10.2.

Table 10.2 The estimation results of the source code similarity for the various systems

Serial number of system	Name of system	Estimates deviation of structure, %	Estimates deviation of size, %	Estimates deviation of complexity, %	Average deviation, %
1	Ant 1.3	22,4	87,7	34,9	48,3
2	Ant 1.4	24,5	82,3	29,2	45,3
3	Ant 1.5	30,3	70,9	26,0	42,4
4	Ant 1.7	46,0	24,7	44,3	38,3
5	Cam 1.2	12,7	51,3	35,7	33,2
6	Cam 1.4	23,6	57,4	33,6	38,2
7	Cam 1.6	12,2	20,6	41,8	24,9
8	Jed 4.3	14,4	42,1	46,7	34,4
9	Luc 2.0	30,9	83,6	45,1	53,2
10	Luc 2.2	31,4	78,9	38,6	49,6
11	Luc 2.4	29,9	68,0	20,9	39,6
12	**Poi 2.5**	**74,6**	**59,0**	**25,4**	**53,0**
13	**Poi 3.0**	**71,8**	**52,5**	**24,5**	**49,6**
14	Pro 45	28,6	40,5	40,5	36,5
15	Pro 451	28,6	51,9	44,5	41,7
16	Pro225	270,1	815,7	52,5	379,4
17	Pro285	350,6	660,9	66,9	359,5
18	Tom 6.0	12,8	19,0	28,7	20,2
19	Xal 2.5	5,1	9,0	5,4	6,5
20	Xal 2.6	0,4	2,2	2,3	1,6
21	**Xal 2.7**	**0,0**	**0,0**	**0,0**	**0,0**
22	Xer 1.4	27,4	56,0	43,9	42,4

The numerical values of the deviations fluctuated ranging from 0 (the basic system 21 taken as a reference point) up to 379% for system 16. It means that system 16 differs from system 21 by 379% by the size, structure, and complexity of the source code. The indexing of the estimation results on the systems names allowed revealing the groups of the systems with close deviations from the basic system. These groups of systems represent various versions of the program. The similarity of the various versions of one system is logical. For example, one version of the system (12) differs from the basic system 21 by 53%. Another version of this system (13) differs from the basic system by 49,6%. Therefore, the source code of versions 12 and 13 differs by 53–49,6 = 3,4%.

The properties similarity of the source code for the various versions of one system received the numerical confirmation in Table 10.2. Thus, the *first similarity principle* from the five offered by us is carried out. Further, if we speak about similarity of the various versions of one system, their similarity is not restricted to the source code similarity. The various versions of one system have the same functional purpose. The functionality of the various versions, as a rule, extends; however, it does not change cardinally. It demonstrates the functional purpose similarity, and *the second similarity principle* is carried out. The various versions of one system are, as a rule, developed in one company by a certain development team. It demonstrates the developers' qualification similarity, and *the third similarity principle* is carried out.

However, in our opinion, the fact of development of the various versions of one system in one company by one developers' group does not guarantee the development processes similarity at all (it is *the fourth similarity principle*). The development process can significantly change from one version to another for different external and internal reasons for the company. The extent of a faultless code use of the own and the third-party development can also change (it is *the fifth similarity principle*).

We consider that the first three similarity principles are carried out with the high probability for the various versions of systems, unlike the fourth and fifth principles, in our case. The observance of these principles is not confirmed by initial data.

Thus, we determined that three out of five similarity principles are true for the studied systems. However our purpose is to estimate the reliability metrics of similar system for the reliability prediction of the new systems. The following question is not answered. What is the reliability degree for the various versions of one similar system, if these versions are (1) the properties similarity of the source code (the first similarity principle is carried out); (2) the similarity of the functional purpose (the second similarity principle is carried out); and (3) the similarity of developers' qualification (the third similarity principle is carried out)? This question is considered in the next section.

10.6.1 *Reliability Metrics and Procedure of Their Research*

We suggest using the following reliability metrics to investigate the reliability similarity of the similar systems:

1. The ratio between faulty modules (FM) and fault-free modules (FFM)
2. The fault localization (FL) in a source code of a system.
3. The fault percentage distributing (FPD) in a source code of a system.
4. A probability of a fault detection (PRFD) in the modules of a system.
5. A modular fault density (MFD) is a faults number in one module of a system.
6. A fault density (FD) is a faults number in 1000 lines a source code.

The feature of the offered metrics consists of the calculation of their estimates depending on the source of code properties. We detail the reliability metrics depending on CA values. We applied the specialized software tool developed for the research objectives to calculate CA values. The simple algorithm of the tool consists of the computation of CA by the formulas (10.6), (7), and in the calculation of the offered metrics on the basis of the known faults number revealed in the verification process. As an example, the estimation result of the reliability metrics for the system 12 from Table 10.2 is shown in Table 10.3.

We visualized the dependences of the reliability metric estimates on the code properties expressed by means of CA after the performance for the other similar

Table 10.3 The estimates of the reliability metrics for system 12

CA	Faults number	LOC	Modules number	FM	FFM	FPD, %	PRFD	MFD	FD
0,20	24	3756	83	22	61	4,84	0,27	1,09	6,39
0,40	196	21,652	150	111	39	39,52	0,74	1,77	9,05
0,60	113	22,848	73	54	19	22,78	0,74	2,09	4,95
0,80	55	17,847	33	26	7	11,09	0,79	2,12	3,08
1,00	12	5611	11	6	5	2,42	0,55	2,00	2,14
1,20	35	10,898	15	13	2	7,06	0,87	2,69	3,21
1,40	6	2542	3	3	0	1,21	1,00	2,00	2,36
1,60	6	14,526	5	2	3	1,21	0,40	3,00	0,41
1,80	2	1390	1	1	0	0,40	1,00	2,00	1,44
2,00	5	1287	1	1	0	1,01	1,00	5,00	3,89
2,20	12	1915	2	2	0	2,42	1,00	6,00	6,27
2,40	1	3728	2	1	1	0,20	0,50	1,00	0,27
2,60	4	1358	1	1	0	0,81	1,00	4,00	2,95
2,80	6	2919	2	2	0	1,21	1,00	3,00	2,06
3,00	1	379	1	1	0	0,20	1,00	1,00	2,64
4,00	7	3446	1	1	0	1,41	1,00	7,00	2,03
5,40	11	3629	1	1	0	2,22	1,00	11,00	3,03
Total	*496*	*119,731*	*385*	*248 64%*	*137 36%*				

systems. We analyzed the similarity of the estimates on each metric on the basis of these dependences. We estimated the similarity degree according to a four-point grading scale; 0 points, the similarity is absent; 1 points, low similarity; 2 points, average similarity; and 3 points, high similarity. After the estimation we defined the resultant assessment of reliability similarity of the systems as the average for the six estimates.

10.6.2 *Research of Reliability Similarity for System Versions*

We need to define to what degree the reliability of the various versions of one system is similar, if these versions have the similar source code (the first similarity principle is carried out) and similar functional purpose (the second similarity principle is carried out) and the versions are developed by specialist of the similar qualification (the third similarity principle is carried out). As a demonstration example, we selected the similar systems 12 and 13 from Table 10.2. The difference of the source code of these systems is insignificant (3,4%). We will analyze the offered metrics for these systems.

The reliability metric 1 is the ratio between faulty modules (FM) and fault-free modules (FFM). We have selected and normalized by the formula (10.6) the metric estimates of the similar systems 12 and 13 from Table 10.2. Further we have calculated CA values by the formula (10.7) and have received the numerical estimates of the reliability metrics after the performance of calculations, groups, and indexing of data. We visualized these estimates on Fig. 10.2. CA values are shown on the abscissa axis; the number of FM and FFM is shown on the ordinate axis. On Fig. 10.2 the dependences between the quantity of FM (red color) and FFM (green color) from CA values are shown.

The calculations have shown that the total ratio between FM and FFM is identical and makes accordingly to 64% and 36% for these systems. The analysis of the diagrams shows the estimations similarity of this metric. It allows drawing the conclusion about the similarity of a ratio between FM and FFM for these similar systems. The practical aspect of this metric consists of the following. The ratio

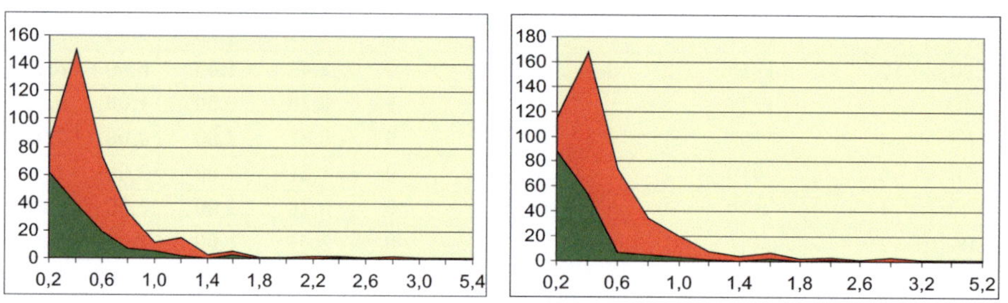

Fig. 10.2 Diagrams of the ratio between FM and FFM for the similar systems 12 and 13

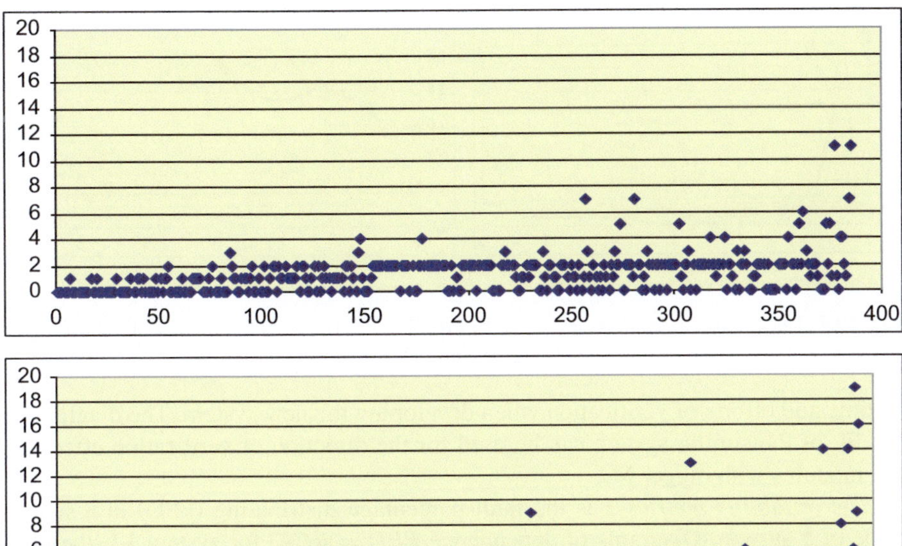

Fig. 10.3 Diagrams of fault localization for similar systems 12 and 13

between FM and FFM shows the fault degree of a system code. If this ratio is higher, it means more time, efforts, and costs the verification process of a code will demand. On the diagram thetop red figure shows the total modules number with the various properties in system. This diagram gives to specialists an accurate account of the properties and reliability of a system code and allows to plan the resources and the processes of the verification and refactoring.

The reliability metrics 2 is the fault localization (FL) in a code. The diagrams of this metric for the similar systems 12 and 13 are shown on Fig. 10.3. The charts show the dependence of the faults number in each module on CA value for this module. The system modules in ascending order of CA values are represented on the abscissa axis. The faulty number in one module of system is represented on the ordinate axis. The points are shown FM (y > 0) and FFM (y = 0) of the system on the chart. The diagrams show the similar number of FFM for both systems. The majority of modules of these similar systems contain 1 or 2 faults. The faults are equally disseminated through a code of these similar systems, except for several modules with a large faults number for the system 13. The total faults amount for system 12 was 496; for system 13 one was 500. These are very close indicators. The analysis of the metrics allows drawing a conclusion about the similarity of the FL in a code of these similar systems. The practical aspect of the metrics consists of the following. The total faults amount of the similar system can be used for the planning

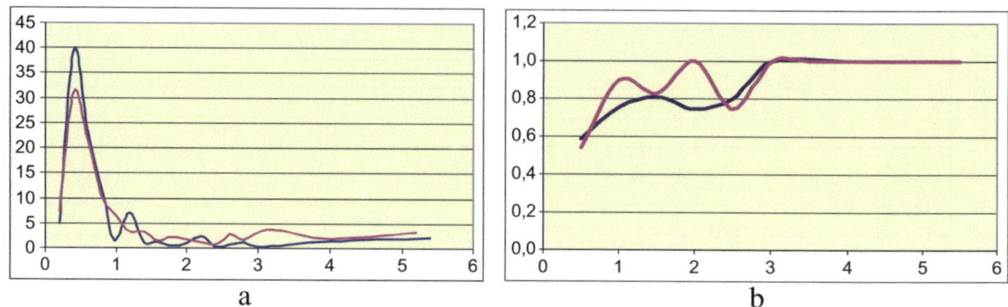

Fig. 10.4 Dependences: (**a**) *FPD = f(CA)*, (**b**) *PRFD = f(CA)* for similar systems 12 and 13

of time and efforts of verification when developing the new system. The diagram of the FL of the similar system can be used for the direction of verification efforts to the modules with bigger FL.

The reliability metrics 3 is the fault percentage distributing (FPD) in a source code of a system. Diagrams of dependence *FPD = f(CA)* for system 12 (the blue curve) and system 13 (the pink curve) are shown on Fig. 10.4a, where the similarity of dependences between a share in % from a total quantity of faults and CA values is presented. The initial and finite coordinates of the curve on y-axis are almost identical for these similar systems. The configuration of dependences is similar. It allows drawing the conclusion about the similarity of FPD in the source code of these similar systems. The practical aspect of the metrics 3 consists of the following. The numerical estimates and dependences of FPD of similar system can be used for the direction of verification efforts to such modules of the new system which contain a bigger amount of faults.

The reliability metrics 4 is the probability of identification of faults (one or several) in the module (probability of fault detection, PRFD). Dependences *PRFD = f(CA)* for system 12 (the blue curve) and system 13 (the pink curve) are shown on Fig. 10.4b.

These dependences have a similar configuration. The initial coordinate of dependences on y-axis is identical for these similar systems, and the value is 0.6. It means that six modules contain faults from each ten modules of the minimum complexity. Both curves contain a direct piece (Y = 1) by CA = 3. It is such a part of a code in which all modules contain faults, one or several. The analyses of dependences allow us to draw the conclusion about the similarity of PRFD in modules for these similar systems. The practical aspect of this metric consists of the following. The PRFD analysis of similar system allows specialists to reveal modules with PRFD = 1 in the new system for their obligatory verification.

We have also paid attention to the following. On Fig. 10.4 we see monotonous and spasmodic pieces of curves. Analyzing various systems, we have defined that the spasmodic curves are the most typical curves for real systems. In our opinion, the jumps of curves can be caused by various objective and subjective factors which are not considered in our research. These factors are the differences of development processes, the use of the earlier verified faultless modules, the insufficient verification efforts, the secondary faults, and some other factors.

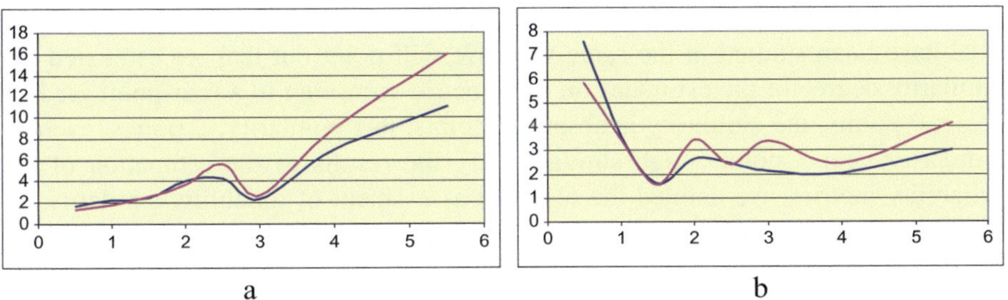

Fig. 10.5 Dependences: (**a**) *MFD = f(CA)* and (**b**) *FD = f(CA)* for similar systems 12 and 13

The reliability metrics 5. The modular fault density (MFD) is number of faults in one module of the system. After MFD estimation we visualized the dependence $MFD = f(CA)$ for system 12 (the blue curve) and system 13 (the pink curve) on Fig. 10.5a. The growing nonlinear dependence is close to exponential curve. The initial coordinates on y-axis and the configuration in the first part of curves coincide. The MFD average value for system 12 is 1.28. The MFD average value for system 13 is 1.13. These are very close indicators. The final coordinates on y-axis and the configuration in the second part of curves differ to some degree. However these differences are explained by the objective error arising because of the small modules number. It reduces the estimation accuracy. The analysis of numerical estimates of this metric and its dependences on code properties allows drawing the conclusion about the similarity of MFD for these similar systems. The practical aspect of this metric consists of the following. The analysis of the MFD for similar system allows specialists to direct the verification efforts to the modules with a large faults number in new system. For example, for systems 12 and 13, these are the modules with $4 < CA < 5.5$. The effort orientation is especially important with limited resources of verification.

The reliability metrics 6 is fault density (FD). FD is faults number on 1000 source code lines. The dependences $FD = f(CA)$ for system 12 (the blue curve) and system 13 (the pink curve) are shown in Fig. 10.5b.

On Fig. 10.5b, we see nonlinear dependences with the jumps. The configuration of these dependences is similar. The initial and final coordinates of dependence on y-axis differ slightly. The average value FD for system 12 is 4.14. The average value FD for system 13 is 3.87. These are close indicators. The configuration of these dependences is explained by the specifics and features of the system development. The analysis of the numerical estimates of this metric and their dependences on code properties allows us to draw the conclusion about the similarity of FD for these similar systems.

Thus, the analysis of numerical estimates of the offered reliability metrics and their dependences on code properties for the similar systems 12 and 13 allows concluding about the high degree of reliability similarity for the studied similar systems.

The reliability metrics of the various versions of similar systems from Table 10.2 have been studied in the same way. We will point out that we estimated the similarity degree of the estimates on each metric according to a four-point grading scale; 0 points, the similarity is absent; 1 points, low similarity; 2 points, average similarity; and 3 points, high similarity. By the results of the estimation of the reliability metrics, we defined the resultant assessment of reliability similarity for the systems as the average one for the six metric estimates. The results received by us are presented in Table 10.4. Among the studied similar systems, we have not found such for which the reliability similarity is absent. We have defined that 79% of the similar systems have high (36%) and average (43%) similarity of the reliability metrics. Twenty-one percent of the similar systems have low similarity. It is necessary to pay attention that the versions of only one similar system have

Table 10.4 The results of reliability similarity estimation for the similar systems

№	Name of system	Reliability metric estimates						Result of reliability similarity estimation
		M 1	M 2	M 3	M 4	M 5	M 6	
1	Poi 2.5	3	3	3	3	3	3	High similarity
	Poi 3.0							
2	Pro 225	2	3	3	3	2	2	High similarity
	Pro 285							
3	Luc 2.0	2	3	2	3	3	2	High similarity
	Luc 2.2							
4	Luc 2.2	3	3	3	2	2	2	High similarity
	Luc 2.4							
5	Luc 2.0	2	2	2	2	2	2	Average similarity
	Luc 2.4							
6	Cam 1.2	3	3	3	2	2	0	Average similarity
	Cam 1.4							
7	Cam 1.4	3	3	3	2	1	1	Average similarity
	Cam 1.6							
8	Cam 1.2	3	3	3	3	2	1	High similarity
	Cam 1.6							
9	Xal 2.5	3	2	3	2	2	3	Average similarity
	Xal 2.6							
10	Xal 2.6	0	0	3	3	3	3	Average similarity
	Xal 2.7							
11	Xal 2.5	0	0	3	3	3	3	Average similarity
	Xal 2.7							
12	Ant 1.3	2	1	0	1	2	0	Low similarity
	Ant 1.4							
13	Ant 1.4	1	2	2	2	2	0	Low similarity
	Ant 1.5							
14	Ant 1.5	2	0	3	1	0	0	Low similarity
	Ant 1.7							

low similarity of the reliability metrics. In our opinion, the low similarity of the reliability is explained by the lack of accounting of two similarity principles in this work. They are 1) differences of the development process for the various versions of this system and 2) use of the fragments of the earlier verified faultless code as the part of the new versions.

The analysis of the values and the dependences of the reliability metrics on the source code properties for the similar systems allow us to draw the conclusion about the reliability estimates similarity of the similar systems and a possibility of the use of such estimates for reliability prediction of new systems.

10.7   Results and Discussion

Thus, the big data analysis technique and concept of similar programs-based assessment of software reliability consists of the procedures: the allocating big data processing of various systems on work servers, data processing performance; the choice of similar system and the selection of her metric data, which is the most informative for estimation of reliability; the reducing of metric data dimensionality by means of the normalization and the convolution; and the estimating software reliability metrics for similar system and use of these metrics for the reliability prediction of new system.

We have three debatable questions after the analysis of the received results. The *first debatable question* is: What metrics are to be used for the accounting of the similarity degree of the development processes and the level of use of a faultless code? The *second debatable question* is: What additional factors, except described by us, influence significantly the reliability of systems? This question consists of the possible extension of the list of the similarity principles.

Moreover, if we sort data of systems according to an average deviation, we will see that the systems of the different functional purpose from the different developers have identical deviations, i.e., are similar in properties of a code. The sorting according to the average deviations unites the systems in the groups which are not connected with the versions. These systems are allocated in Table 10.5.

There is *the third debatable question*. How similar is the reliability of such systems? Our further research will be connected with the search of answers to these questions.

Table 10.5 The estimation results of the source code similarity for the various systems

Serial number of system	Name of system	Estimates deviation of structure, %	Estimates deviation of size, %	Estimates deviation of complexity, %	Average deviation, %
1	Xal 2.7	0,0	0,0	0,0	0,0
2	Xal 2.6	0,4	2,2	2,3	1,6
3	Xal 2.5	5,1	9,0	5,4	6,5
4	Tom 6.0	12,8	19,0	28,7	20,2
5	Cam 1.6	12,2	20,6	41,8	24,9
6	Cam 1.2	12,7	51,3	35,7	33,2
7	Jed 4.3	14,4	42,1	46,7	34,4
8	Pro 45	28,6	40,5	40,5	36,5
9	**Cam 1.4**	**23,6**	**57,4**	**33,6**	**38,2**
10	**Ant 1.7**	**46,0**	**24,7**	**44,3**	**38,3**
11	Luc 2.4	29,9	68,0	20,9	39,6
12	Pro 45	28,6	51,9	44,5	41,7
13	**Ant 1.5**	**30,3**	**70,9**	**26,0**	**42,4**
14	**Xer 1.4**	**27,4**	**56,0**	**43,9**	**42,4**
15	Ant 1.4	24,5	82,3	29,2	45,3
16	Ant 1.3	22,4	87,7	34,9	48,3
17	**Luc 2.2**	**31,4**	**78,9**	**38,6**	**49,6**
18	**Poi 3.0**	**71,8**	**52,5**	**24,5**	**49,6**
19	**Poi 2.5**	**74,6**	**59,0**	**25,4**	**53,0**
20	**Luc 2.0**	**30,9**	**83,6**	**45,1**	**53,2**
21	Pro285.0	350,6	660,9	66,9	359,5
22	Pro225.0	270,1	815,7	52,5	379,4

10.8 Conclusion and Future Work

The conducted research allowed formulating a number of results and the discussion questions not clarified yet.

1. The motivation of the use of the big data analytics for the reliability increase of program systems is formulated. This motivation consists of the reduction of the billion losses as a result of the faults and failures of systems.
2. The review of the scientific and technical literature in the field of the big data analysis is done. It is offered to adapt the models and methods of big data analysis for the tasks of the estimation, prediction, and increase of software reliability.
3. It is offered to use a similar system for the reliability prediction of the new system. The concept of similar programs on the basis of five principles is formulated. The first principle is based on the size, structure, and complexity metrics. The search results of the similar program based on the first principle are represented. The system has been identified with minimum 5,1%, maximum 9%, and average 6,5% relative deviation from metrical rates among the 20 explored

systems. The obtained results confirm the allegation that the systems with known reliability indexes similar to a new system under development may be found from a great quantity of the experimental data kept in the big data storage to predict software reliability.

4. The adaptation of the MapReduce model for the search and estimation of the similar system properties is offered. The procedure of the selection and the convolution of metrics in the uniform combined assessment of the source code properties are offered. The use of the combined assessment significantly reduces the data size and simplifies the visualization and the analysis of system properties.

5. It is offered to estimate the reliability of a similar system depending on the properties of a code by means of the following detailed metrics: ratio between faulty modules and fault-free modules, fault localization, fault percentage distributing, probability of fault detection, modular fault density, and fault density. The analysis, the visualization, and the interpreting for the offered reliability metrics are carried out.

6. The analysis of the studied similar systems has shown that 79% of systems have high (36%) and average (43%) similarity of the reliability metric estimates. Twenty-one percent of similar systems have low similarity of reliability metric estimates. The lack of the similarity of the reliability metric estimates among the studied similar systems was not revealed.

7. The received results allow us to draw a conclusion about the similarity of the reliability metric estimates of the similar systems and a possibility of the use of these estimates for the reliability prediction of new systems. The developers of systems can use the predictive reliability estimations for the resource management of the verification and refactoring processes. These activities will provide the reliability increase of new program systems under condition of cutting costs for their development.

The further work will be directed to the research of the similarity principles which impact on the reliability, but haven't been considered in our research, in particular, development (or forming) of matrixes of similarity properties for different systems based on all similarity principle and reliability for these systems based on their experience or operation results and comparing and correlation analysis of these matrixes to improve reliability prediction quality.

References

1. Mayevsky, D. A. (2013). *A new approach to software reliability. Lecture notes in Computer Science. Software engineering for resilient systems* (Vol. 8166, pp. 156–168). Berlin: Springer.
2. Yakovyna, V., Fedasyuk, D., Nytrebych, O., Parfenyuk, I., & Matselyukh, V. (2014). Software reliability assessment using high-order Markov chains. *International Journal of Engineering Science Invention, 3*(7), 1–6.

3. Yakovyna, V. S. (2013). *Influence of RBF neural network input layer parameters on software reliability prediction.* 4-th International Conference on Inductive Modelling, Kyiv, pp. 344–347.
4. Maevsky, D. A., Yaremchuk, S. A., & Shapa, L. N. (2014). A method of a priori software reliability evaluation. *Reliability: Theory & Applications, 9*(1, 31):64–72. Access mode: http://www.gnedenko-forum.org/Journal/2014_1.html
5. Yaremchuk, S. A., & Maevsky, D. A.. (2014). The software reliability increase method. *Studies in Sociology of Science,* 5(2):89–95. Access mode http://www.cscanada.net/index.php/sss/article/view/4845
6. Kharchenko, V. S., Sklar, V. V., & Tarasyuk, O. M. (2004). *Methods for modeling and evaluation of the quality and reliability of the software.* Kharkov: Nat. Aerospace. Univ."KhAI". 159 p.
7. Kharchenko, V. S., Tarasyuk, O. M., & Sklyar, V. V. (2002). The method of software reliability y growth models choice using assumptions matrix. In *Proceedings of 26-th Annual International Computer Software and Applications Conference* (pp. 541–546). Oxford, England: COMPSAC.
8. Carrozza, G., Pietrantuono, R., & Russo, S. (2012). Fault analysis in mission-critical software systems: A detailed investigation. *Journal of Software: Evolution and Process, 2,* 1–28. https://doi.org/10.1002/smr.
9. Manyika, J., et al. (2011). *Big Data: The next frontier for innovation, competition, and productivity.* McKinsey Global Institute. https://bigdatawg.nist.gov/pdf/MGI_big_data_full_report.pdf
10. Capgemini (2015). *Big & fast data: The rise of insight-driven business.* http://www.capgemini.com/insights-data
11. A ComputerWeekly buyer's guide to data management. (2017). http://www.computerweekly.com
12. Big data poses weighty challenges for data integration best practices. Information management handbook. (2017). http://www.techtarget.com/news
13. Dijcks, J.-P. (2013). *Oracle: Big Data for the enterprise.* http://www.oracle.com
14. DLA Piper & BCG. (2015). *Earning consumer trust in Big Data: A European perspective.* Carol Umhoefer, Jonathan Rofé, Stéphane Lemarchand. DLA Piper, Elias Baltassis, François Stragier, Nicolas Telle – The Boston Consulting Group. pp. 20.
15. Bridget Botelho at all. (2016). *Big Data warriors formulate winning analytics strategies.* E-publication. TechTarget Inc., www.techtarget.com
16. Gartner: Seven Best Practices for Your Big Data Analytics Projects. (2015).
17. Best Practices for a Successful Big Data Journey. (2017). Datameer, Inc. http://www.bitpipe.com/fulfillment/1502116404_933
18. Meeker, W. Q., & Hong, Y. (2014). Reliability meets Big Data: Opportunities and challenges. *Quality Engineering, 26*(1), 102–116., Taylor & Francis Group.
19. Zenmin, L. (2014). *Using Data Mining Techniques to improve software reliability.* Dissertation for the degree of Doctor of Philosophy in Computer Science, p. 153. https://www.researchgate.net/publication/32964724/
20. Kharchenko, V., & Yaremchuk, S. (2017). Technology Oriented assessment of software reliability: Big Data based search of similar programs. In *Proceedings of the 13th International Conference on ICT in Education, research and industrial applications* (pp. 686–698). Integration, Harmonization and Knowledge Transfer. Workshop TheRMIT.
21. Leskovec, J., Rajaraman, A., & Jeffey, D. (2014). *Mining of massive datasets.* Stanford University, Milliway Labs., p. 495.
22. Lammel, R. (2007). *Google's MapReduce programming model—Revisited.* Data Programmability Team Microsoft Corp. Redmond, WA, USA, pp. 1–42. https://userpages.uni-koblenz.de/ laemmel/MapReduce/paper.pdf
23. Belazzougui, D., Botelho, F. C., Dietzfelbinger, M. (2009). Hash, displace, and compress (pp. 1–17). Berlin/Heidelberg: Springer. http://cmph.sourceforge.net/papers/esa09.pdf

24. Dean, J., & Ghemawat, S. (2004). *MapReduce: Simplified data processing on large clusters.* OSDI'04: Sixth Symposium on Operating System Design and Implementation, San Francisco, CA, pp. 1–13. https://research.google.com/archive/mapreduce.html
25. Tera-PROMISE Home. (2017). http://openscience.us/repo/fault/ck/
26. NASA'S DATA PORTAL. (2017). https://data.nasa.gov
27. Software Testing and Static Code Analysis. (2017). http://www.coverity.com
28. Topcoder | Deliver Faster through Crowdsourcing. https://www.topcoder.com (2017)
29. Chidamber, S., & Kemerer, C. (1994). A metrics suite for object-oriented design. *IEEE Transactions on Software Engineering, 20*(6), 476–493.

Index

© Springer International Publishing AG, part of Springer Nature 2018

M. M. Alani et al. (eds.), *Applications of Big Data Analytics*,
https://doi.org/10.1007/978-3-319-76472-6

 MIX
Papier aus verantwortungsvollen Quellen
Paper from responsible sources
FSC® C105338
www.fsc.org

If you have any concerns about our products,
you can contact us on
ProductSafety@springernature.com

In case Publisher is established outside the EU,
the EU authorized representative is:
**Springer Nature Customer Service Center GmbH
Europaplatz 3, 69115 Heidelberg, Germany**

Printed by Libri Plureos GmbH
in Hamburg, Germany